Animal Bioethics
Old Dilemmas and New Challenges

Edited by

Zoran Todorović and Siniša Đurašević

Animal Bioethics: Old Dilemmas and New Challenges
Edited by Zoran Todorović and Siniša Đurašević

This book first published 2022
Ethics International Press Ltd, UK
British Library Cataloguing in Publication Data
A catalogue record for this book is available from the British Library
Copyright © 2022 by Ethics International Press Limited

Print Book ISBN: 978-1-80441-016-5
eBook ISBN: 978-1-80441-017-2

CONTENTS

PRAISE FOR ANIMAL BIOETHICS

The new publication entitled "Animal Bioethics - Old Dilemmas and New Challenges" edited by Prof. Zoran Todorovic (School of Medicine, University of Belgrade, Serbia) and Prof. Sinisa Djurasevic (Faculty of Biology, University of Belgrade, Serbia) covers a wide range of topics related to animal experimentation. The authors explore the studied issues from different angles including ancient philosophical views as well as the current animal welfare science approach or relevant EU legal regulations.

To ensure the high quality of each chapter, the editors invited experts in the areas of ethics, bioethics, physiology, neurology, toxicology, pharmacology, animal welfare and legislation to provide specific texts within their area of expertise. Overall, the book is a comprehensive source of information including ethical evaluations that should not be missed by those interested in animal experimentation, whether they are researchers or people arguing for the needs of laboratory animals.

Prof. Ing. Eva Voslarova, Ph.D.
Head of Animal Protection and Welfare Unit, Department of Animal Protection and Welfare and Veterinary Public Health, University of Veterinary Sciences Brno, Czech Republic

The use of animals for scientific purposes was, and still is, a subject of debate to its true usefulness. Although there is a long list of scientific achievements which improved, one way or another, the quality of life not only for humans but also for animals, the question if we need to use animals for research purposes remains unanswered.

Nowadays, when the interest of the community for the environment is even greater, the use of animals for research purposes is no longer a controversial issue but an issue that seeks clear answers. The first answer that can certainly be given is that this use has a clear ethical background that must be taken seriously under consideration. The use of animals in experiments is not a *de facto* human right. It is a need that should be based on the moral obligation of humans to respect non-human animals. It is a necessity that should only be realized when humans, animals or the environment benefit from the obtained results. It is an act that is covered by legal requirements, international scientific and ethical rules, ensuring the proper care and use of laboratory animals as well as the quality of the obtained experimental results.

Although thousands of pages have been written on this subject, there is always room for more. This is exactly what the book "*Animal Bioethics: Old Dilemmas and New Challenges*", edited by Professors Zoran Todorovic and Sinisa Djurasevic aspires to achieve. Through an objective look, the authors try to approach the subject holistically, covering all scientific, legal and bioethical aspects. They do not want to influence the reader for or against the use of animals. What they want is to inform about the new scientific achievements, the new challenges and the up-to-date ethical concerns that regulate the use of animals. I do not know whether, after completing the study of the book, the reader will finally be able to answer the dilemma of using animals for research purposes or not. What I can assure you is that by completing the study of this book the reader will have more knowledge to ponder.

Nikolaos Kostomitsopoulos, DVM, PhD
Centre of Experimental Surgery, Biomedical Research Foundation, Academy of Athens, Athens, GREECE

Some ancient religions of the East have long warned that animal life and suffering are worthy of human attention and consideration. In the West, however, awareness came only in the first half of the 19th century, at about the same time in England and among German pietists. They were the desperate, lonely cries of the first societies that advocated for mercy.

Two hundred years later, however, we are witnessing a very strong initiative to protect and expand animal rights: and although we are talking about hunting animals, using animals in circuses, eating animal meat, society's primary focus is on the use of animals in scientific experiments. The argument against their use is clear: if the stronger is allowed everything just because it is stronger, will we stop at animals or will we switch to minorities and people with disabilities? There is no progress in the logic of argumentation: the view of the majority is summed up in an aphorism signed under a photo of anti-vivisection protesters - "Animal experiments have allowed you to demonstrate twenty years longer..." In practice, however, the shifts are very visible: since the famous '3R', launched at the beginning of the second half of the 20th century, to the massive refusal of ethics commissions to issue permits for experiments on non-human primates.

And while public discourse - sometimes in national parliaments - is moving on to sexual intercourse with pets, plant rights, etc., tens of thousands of diligent scientists in laboratories all around the world remain confused and even frightened by modern trends in legislation and ethics. Addressing them, the book "Animal Bioethics: Old Dilemmas and New Challenges," edited by Zoran Todorovic and Sinisa Djurasevic, comes as more than a good guide to help navigate the forest of laws, directives, codes, and attitudes, setting clear limits but also allowing good practice to evolve without stopping the progress of science.

Prof. Amir Muzur, MD, PhD
University of Rijeka, Faculty of Medicine, Department of Social Sciences and Medical Humanities, and Faculty of Health Studies, Department of Public Health, Rijeka, CROATIA

PREFACE

This book deals with various aspects of animal bioethics, from philosophical foundations to genetically modified organisms and their impact on the environment. Such a concept corresponds to the holistic approach in bioethics of Fritz Jahr and Van Rensselaer Potter, which observes the integrity of man's relationship with all forms of life and the environment. A background of monistic and dualistic concepts of the human-animal relationship is depicted as well. Then experimental models in drug development and pain testing are analyzed, and the translational aspect of in vivo experiments. A particular chapter is dedicated to neuroethics, taking into account the importance of animal experiments for examining brain function. Finally, an overview of modern legislation related to animal experiments is given, the ethical basis of the principles of Good Laboratory Practice is assessed, and the importance of animal bioethics for writing scientific projects is shown.

There are just a few publications dealing with animal bioethics. It is worth mentioning the book under a similar title published fifteen years ago: Marie M et al., Eds. Animal Bioethics: Principles and Methods of Teaching (1st Edition). Wageningen, NL: Wageningen Academic Publishers, 2005. Our book will predominantly cover topics not described or omitted from this publication (see the short description above). Other available literature does not have such an integrative approach. The authors are experts in various aspects of animal bioethics, from philosophy, medicine, and biology to veterinary issues and environmental protection. The unique quality of our publication is reflected in the detailed presentation of modern legislation in this area and the principles of drafting scientific projects in which animal bioethics has an important place.

Our publication is interesting for the broadest audience, from students of biomedicine and related fields to scientific researchers and members of organizations for the protection of animal rights and welfare. Some chapters will certainly be interesting to media representatives and the general public, not to mention employees in the state administration as potential stakeholders. Editors and authors have extensive experience in animal bioethics studies. They have successfully established a national system for protecting animal welfare and served as lecturers on behalf of the RSPCA. Additionally, they have a rich scientific background in the field. The editors and authors have participated for over a decade in a joint master's program in research ethics with the Icahn Faculty of Medicine at Mount Sinai. Last but not least, one of the editors of this book has already served as the first editor of an international monograph entitled: "Bioethics and Pharmacology: Ethics in Preclinical and Clinical Drug Development" (Kerala, India: Transworld Research Network, 2012).

Belgrade, May 2022
Zoran Todorović
Siniša Đurašević

ANIMAL (BIO)ETHICS – A PHILOSOPHICAL BACKGROUND

Dejan Donev

Full professor of Ethics, History of Ethics, Ecoethics and Bioethics ss Cyril and Methodius University in Skopje, Faculty of Philosophy, Institute of Philosophy, Skopje, N. Macedonia

Abstract

Today, we are still dealing with the unresolved question about the relationship between humans and animals, which belong to distinctly and significantly different ontological stages. Can this ontological differentiation, which imposes certain insurmountable limits of argumentation in favour of a behaviour, guided by moral rules, concerning animals and the very thought of their rights in general, be considered sufficient? Or, in the modern ethical discussion, we should require an adaptive reorientation of the argument, if it refers to the normative regulation of our behaviour towards animals? Is it possible to create and apply animal bioethics?

Keywords: bioethics, animal (bio)ethics, utilitarianism, humans, animals, moral status

Introduction - Do animals have a moral status!?

Issues related to animal rights and our duties following their *moral status* have become an essential part of almost all major ethical debates in the last few decades, representing one of the most current areas of ethics research. This is because the "human beings relation on the planet to living non-human beings, animals, is characterised by their apparent superiority. Thanks to their overall abilities and potentials, human beings have become masters of the planet. Their *dominant* planetary position raised the question of the value regulation of their behaviour towards non-human beings, animals as *lower and subordinate species*. (...) Throughout the history of civilisation, people have often treated their superiority as an implicit or explicit authority for complete submission to non-human beings, animals in relation to human demands, interests and needs, with the behaviour towards them being determined as *morally indifferent*." (1)

Thus, in the 1960s, the global animal rights movement was create[1], and the role of philosophy in developing the theoretical framework and forcing intellectual debates about our treatment of animals[2] was crucial in addressing this issue in its full significance - moral justification of current practices and regulation of normative issues regarding the attitude towards animals, in general[3]. Peter Singer, one of the most important representatives of this movement[4], commented on the role of philosophy, which joined this movement as a science in the 70s of the last century, saying that: "philosophers were not the mother of the movement, but they did ease its passage into the world and – who knows – may have prevented it being stillborn". (2) Then, together with Tom Regan (3) and Klaus Michael Meyer-Abich (4), the primary thoughts that are representative of the current discussion on the new thinking and regulation of the relationship between humans and animals, they formulated in the following paragraphs:

1. animals are beings that are capable of suffering[5], with their interests and needs that are similar to the basic needs of people;
2. if there is such similarity, the principle of equality requires that the interests of animals are respected as well as the similar interests of humans;

---

[1] If we go deeper we can find that these kinds of initiatives are already mentioned in the Pietist's works from XVIII century.

[2] Although linguistically and logically disputable, the correct term should be "non-human animals". The term "non-human animals" is used to shed light on the often overlooked fact that humans are also animals. For the rest of this text, we will generally stick to such uses, except when the sources we use relate to the more traditional human and animal dichotomy.

[3] More detailed see: Sirilnik, B. & Fontene, E de. & Singer, P. *I životinje imaju prava*. Novi Sad: Akademska knjiga, 2018, str. 15-17.

[4] but also the founder of the Animal Liberation Movement, along with Jacques Cosnier and Hubert Montagner as early as the 1970's, shortly after the formation of the Oxford Group of Richard Ryder which defined the great principles of animalistic ethics in the collection entitled *Animals, Men and Morals* on Roslind and Stanley Godlovitch and John Harris.

[5] In a well-known passage, which represents a departure from the mainstream of Western philosophy, Bentham says the following: „The day *may* come when the rest of the animal creation may acquire those rights which never could have been withholden from them but by the hand of tyranny. The French have already discovered that the blackness of the skin is no reason why a human being should be abandoned without redress to the caprice of a tormentor. It may one day come to be recognized that the number of the legs, the villosity of the skin, or the termination of the *os sacrum* are reasons equally insufficient for abandoning a sensitive being to the same fate. What else is it that should trace the insuperable line? Is it the faculty of reason, or perhaps the faculty of discourse? But a full-grown horse or dog is beyond comparison a more rational, as well as a more conversable animal, than an infant of a day, or a week or even a month, old. But suppose they were otherwise, what would it avail? The question is not, Can they *reason*? nor Can they *talk*? but, Can they *suffer*?". Bentham J. *Introduction to the Principles of Morals and Legislation*. New York: Prometheus Books, Amherst, 1988, p. 311.

3. animals have their value, which for some (Singer and Regan) stems from their consciousness, while others (Meyer-Abich) attribute additional importance to the affinity of animals and humans.

Animals should, furthermore, be guaranteed the fundamental "right" to life appropriate to their species, the view that is based on the parts of the fourth and fifth articles of the "Universal Declaration of Animal Rights": "wild animals have the right to live and reproduce in freedom their own natural environment ... Any animal which is dependent on man has the right to proper sustenance and care". (5)

Nevertheless, the unresolved question remains about the relationship between humans and animals, which belong to distinctly and significantly different ontological stages. Can this ontological differentiation, which imposes certain insurmountable limits of argumentation in favour of a behaviour, guided by moral rules, in relation to animals and the very thought of their rights in general, be considered sufficient or in the modern ethical discussion should require an adaptive reorientation of the argument, if it refers to the normative regulation of our behaviour towards animals?[6] Is it possible to create and apply animal bioethics? (6)

The Western European tradition of thought and the moral status of animals

Philosophy is responsible for many of our views regarding the natural world, i.e. philosophical thought that from the very beginning influenced our absolute inclination towards anthropocentric ethics, regardless of the consequences of such ideas we face, especially in the last few decades.[7]

The dominant anthropocentric philosophical theories represent the belief that only man, a self-conscious being with the ability to act morally, and thus

---

[6] Futher consult: Protopapadakis, D. E. "Animal Rights, or Just Human Wrongs?", in: *Animal Ethics Past and Present Perspectives*, Protopapadakis D. E. (ed.). Berlin: Logos Verlag Berlin GmbH, 2012, pp. 279-291.

[7] For centuries, the position and status of animals has been neglected or even in no way had its place on earth. The anthropocentric nature of this view of the world was an important reason why our natural-technical civilization did not develop in harmony with nature, but much more often in opposition to it. Elements, minerals, plants and animals are treated, not as "partners", but as resources that man uses indefinitely, so the wildlife in the jungle has fallen victim to large game hunters, and in rich countries the victim of industrial meat production.

the only autonomous being, can have moral status. Namely, "the dignity of a certain individual is considered from the perspective of the reason of one's nature, and such nature is only attributed to man. Only he is free from the realm of goals, while non-human living beings are connected by the bonds and relationships that exist in nature. Only man is aware of himself and can distance himself from himself in favour of higher goals, relativise his interests, all the way to self-surrender. This, as a moral being, gives him an absolute status that establishes his unique dignity, which in turn gives him the right not to be "enslaved" by anyone and, as a moral being, not to be deprived of his own goals. From this unique human dignity arises his unique rights" (7), guaranteed by the UN Declaration of Human Rights from 1948. (8)

Contrary to this, non-anthropocentric theories believe that there is no strict hierarchy between beings in nature. All differences between humans and animals should not be established ontologically but on a biological basis. (9) This means that the predominantly anthropocentric picture of the world and the derived relation of man toward nature and animals in the last few decades has been called into question by the non-anthropocentric expansion of ethics and the publication of (bio)ethical search for a new foundation of the relationship between humans and animals. In this context, Singer speaks of animals as "persons", while Regan speaks of them as "subjects of life" and Mayer-Abich of their dignity, from which they derive animal rights, appropriate treatment of their species, as well as for the protection of their lives, because it is forbidden to kill them for our diet. (10)

These two opposing lines of thinking about the differences or similarities between humans and animals characterise the Western philosophical-religious tradition[8]. The same, in general, which is based on two sources, shows less respect for animals and the rest of the natural world than the Eastern traditions due to the assumed absolute superiority of man over nature[9].

---

[8] The Eastern philosophical-religious tradition is in many ways opposed to Western cultural, religious and philosophical views and nurtures the idea of holiness and enthusiasm for all forms of life, not just human.

[9] which led to a kind of crisis of man in the scientific and technical age when the imperative for the preservation of nature and natural biodiversity is increasingly imposed in relation to the devastating action of man over the natural world.

The first source primarily refers to the nature of the relationship between man and animals as described in the Old Testament Bible, the essential Christian religious-historical document. It gives us an idea of the world in which man was created according to the image of God and is thus destined to rule over all other living things on earth. The divine origin of man is not explained only by his absolute supremacy over other earthly beings. Still, it is also reflected that man is only blessed through the unity of body and immortal soul. The soul is considered a valuable component of humanity, all the more so because it enables moral thinking and action. In contrast, bodily beings, animals, are considered the most primitive level of existence. This view is maintained by the Old Testament stories of animal sacrifice. That is, animals and the rest of the natural world, according to this source, the Christian philosophical-religious doctrine, exist only for human benefit because human beings "are the only morally important members of this world. Nature itself has no intrinsic value, so the destruction of plants and animals cannot be wrong unless the destruction does not harm human beings" (11).

The most influential medieval Christian thinkers, Aurelius Augustine and Thomas Aquinas, are on this first line of thinking. They emphasised that the lack of reason in animals justifies their subordinate role. Namely, Thomas Aquinas, in his philosophical teaching, states the thesis that we are not obliged to show any kindness, mercy or compassion to animals because animals cannot experience any benefit due to their irrational nature, so their value is judged solely by how much they are suitable for human use: "Dumb animals... are devoid of the life of reason... they are naturally enslaved and accommodated to the uses of others" (12). Although it is acknowledged that irrational animals are sensitive to pain, this is not considered a sufficient reason for them to be respected in the same way as human beings. The compassion that can be felt towards suffering animals stems from the moral affection of people, so the individual who does good to people will take care of his property, i.e., the animals it owns.

The second source derives from ancient philosophy[10]. It is most marked by Aristotle's philosophical teaching (13), to which animals, according

---

[10] Which is posed on Aristotle's paragraph from the *Politics* (1256b15-22), emphasized as a paradigm of the leading western tradition and its unquestionable anthropocentrism: „and that plants exist for the sake of animals and the other animals for the good of man, the domestic species both for his service and for his food, and if not all at all events most of the wild ones for the sake of his food and of his supplies of other kinds, in order that they may furnish him both with clothing and with other appliances. If therefore nature makes nothing without purpose or in vain, it follows that nature has made all the animals for the sake of men".

to the natural hierarchy, are under humans because they possess only senses, but not reason and beliefs (14), i.e. rational soul so that man can use them as his resources. However, unlike the animal soul, the human soul is characterised by reason whose products can be communicated through language, thoughts, and reason.

Animals have a sensitive soul but not rational mechanisms for communicating pain, suffering or pleasure, so we should not treat them like any other human. Hence, it can be concluded that animals precisely because of a lack of reason cannot participate in the ethical and political sphere: "our action is outside the sphere of righteous action, so there is no friendship because we have nothing in common" (15). But human primacy over animals can also have a benevolent character. Namely, as Martha Nussbaum states, "Aristotle's great contribution is the idea that each species has its form of life and that it is equipped for life under certain circumstances. The same thing means that the good is different for every animal, and every animal has a purpose in itself and is a measure of its success. (...) Aristotle's claim from *Politics* that animals exist because of humans is contradicted by hundreds of statements from biological writings which suggest that each animal has its purpose" (16).

Exposed ideas about the position of animals in the human world were developed. They later reached their peak with the development of modern Western philosophical thought in the 17th century, reflecting the still dominant influence of Christianity. Anxiety towards animals and the assumption that justifies cruelty to them (that they cannot feel pain) is most prominent in the most important representative of the philosophy of rationalism, Rene Descartes. The differentiation of the overall battle into thinking and stretching stems from Descartes' mechanistic understanding of nature under the influence of the science of mechanics, in which animals as stretching are deprived not only of the thinking and rational aspect of battle but also of feeling. Only human beings, as thinkers, have consciousness in their body, the unique ability of language and innovative behaviour.

For philosophers of later eras such as Hume and Schopenhauer, above all, the dominance of reason in the autonomy of a being has maintained the same direction in philosophical considerations on the question of the moral status of animals. "Animals have no rights," Hume said, although "the laws of humanity do oblige us to treat these creatures with care" (17). This careful treatment follows as an implication from the view that the compassion we can feel for animals can be a source of moral thought, but that does not mean that cruel treatment of animals can be considered as a matter of justice because justice is a moral attitude that refers to equals in force and rights. So, the righteous deed, i.e. equal respect for interests, applies only to human interests. In this context, under the influence of Indian philosophy, Schopenhauer integrated many ideas from Eastern philosophy into his philosophical teaching. While rejecting reason and self-awareness as necessary for the assumption of the moral status of a particular being, he builds the doctrine of the moral treatment of animals around the ethics of compassion. Moreover, the greater intelligence of human beings also increases their ability to suffer, thus justifying the increased moral concern for human suffering" (18).

In modern philosophy, many philosophers, especially Kant, advocated equal respect for interests and autonomy as a fundamental moral principle (19). Still, the same principle did not extend beyond the limits of its kind. Moving forward makes the founder of modern utilitarianism, Jeremy Bentham, who by "autonomy" means "the ability to choose and make one's own decisions, and hence to act accordingly. Rational and self-conscious beings, it seems, have that ability. In contrast, beings who cannot consider the alternatives that are opened to them cannot choose in the required sense and therefore cannot be autonomous. In particular, only a being who can comprehend the difference between dying and living can be autonomous and choose to live" (20).

From the above[11], it can be concluded that the direction in which such philosophical thinking goes is to act rationally, meaning to act morally. To act morally means to have moral significance concerning one's right and not for the interests of others. The ethical attitude according to which we have a direct moral duty only to other moral subjects supports Kant's ethics

---

[11] It can be said for sure that the history of the ideas about the moral status of the animals doesn't stop here. On the contrary! Here we can include also the ideas, thoughts and efforts of Fritz Jahr, Ignaz Bregenzer, Mark H. Bernstein or even Corey Lee Wrenn. In this sense this chapter will be richer, but also very long. So, we decide to leave some of them for in-depth analysis in the future occasion.

of indirect duties. For example, indirect moral duties toward animals, such as compassion and goodwill, are based on personal interest because our good treatment of animals also strengthens our "goodwill" relationship with humans. But Peter Singer criticises the denial of moral importance to other species regardless of the purpose of rational beings, which is a fundamental premise of Kant's argument. He explains that "it may be true that kindness to humans and other animals goes hand in hand; but, nevertheless, Aquinas and Kant argued that this was the real reason one should be kind to animals, which is a total specistic position" (21).

And so, until the 20th century, it was common to think that animals were immature creatures and undeserved of our compassion. In summary, Paola Cavalieri[12] speaks of "a twenty-century philosophical tradition aimed at excluding members of species other than ours, outside the ethical domain" (22). After Aristotle and Thomas Aquinas, she cites Kant as a modern follower of the most persistent and widely accepted thesis of animal relations in all Western culture. By categorising animals as "things" to which we cannot have direct duties: "animals, as pure means have a moral status of zero-order, i.e., they are excluded from the moral community" (23).

Although these are attitudes that have been slowly losing importance in the last few decades as too harsh and unfounded[13], in general, nevertheless, it can be freely said that these are views that are pretty suitable for promoting those attitudes that led to the monstrous destruction of nature by the modern technological society that continues today with increasingly pronounced savagery.

---

[12] An Italian philosopher who, together with Peter Singer, founded the Great Apes Project, a movement created in 1996 that aims to extend to the great apes the three, not all, rights that until then were reserved only for man: the right to life, the right to individual protection and the right for respecting the physical integrity (prohibition of torture). See in more detail: Sirilnik B. & de Fontene E. & Singer P. *I životinje imaju prava*. Novi Sad: Akademska knjiga, 2018, str. 17.

[13] As Prof. Ćović states, it is believed that most of the discussions about human responsibility for non-human living beings take place today within the so-called animal ethics, which task is to determine the "moral status of animals", and within the advocacy of "rights of animals". He adds that the "absurdist method of the speciesistic leveling" has been established within the mentioned framework, which occurs in two forms, "as an Aesopian approach of "leveling upwards", which consists in anthropomorphic attribution to non-human living beings of specifically human properties and categories such as dignity, moral status, rights, etc., and as a Singer's approach of "leveling down", consisting of zoomorphic reduction of specifically human traits and categories. Both procedures have the same goal - to level the gap between man and other living beings with the ability to feel, starting from the mistaken assumption that this is a good way to develop moral considerations and legal obligations toward non-human members of the sensitive community." Ćović A., „Biotička zajednica kao temelj odgovornosti za ne-ljudska živa bića", in *Od nove medicinske etike do integrativne bioetike*, Ćović A. & Gosić N. & Tomašević L. (eds.). Zagreb: Pergamena / Hrvatsko bioetičko društvo, 2009, str. 36-37.

Current (bio)ethical theories about the moral status of animals

Historically, various views and considerations have been encouraged, i.e. various theories about the attitude towards animals have been created. Each of these theories tries through ethics and philosophy to clarify the person's attitude behind that attitude by directly asking the question about the moral status of animals. This is because, in order to be able to attribute and/or recognise moral status to certain beings as a kind of opportunity for proper consideration of objects with direct moral significance, we need to take a particular moral view that they have "a kind of importance as beings, that they have their moral significance, importance" (24). Here, too, when we consider issues related to moral values, others are often involved because they are in some relationship, which says that to have a moral status means "to be an entity concerning which others (living beings) have, or they may have moral obligations" (25). When an entity is given moral status, it does oblige us not to behave as we wish, but it must be well considered whether the same entity has its interests, desires, beliefs and the like. It should be borne in mind that the rules are not binding us on such action, but the intrinsic value of the entity itself, the meaning it has in itself, according to some "own right" (26).

When we talk about animals, the question of moral status, i.e. the value in itself of the same as an entity, covers a number of different criteria, which first refer to various theories of moral status. According to several authors, "rationality is the main criterion for moral action (Aristotle, Kant), while others will be based on the Christian tradition according to which moral status is under the principle of sanctity of life, and it belongs exclusively to human species, third, most often utilitarians will focus on sensitivity as a criterion for moral status that morally binds not only people but also all other living beings who may experience pain, suffering or other mental states, fourth, ecofeminists, to the ability for compassion and care (27) ... So, there are numerous different approaches, i.e. a wide range of theories that try to explain the relevant positions for our relationship with animals and the living world of the earth (28), of which, as the most serious and mature in their positions, we single out the following few[14]:

---

[14] Although others can be found in various representative works. So for example, M.A. Warren offers the following: The Moral Agency Theory, The Genetic Humanity Theory, The Sentience Theory, The Organic Life Theory and Two Relationship-based Theories. Warren M. A., "Moral status", in: *A Companion to Applied Ethics*, Frey G. R. and Wellman H. C. (eds.). Oxford: Blackwell Publishing Ltd., 2003, pp. 440-445.

- **utilitarianism or ethical humanism**, which in the behaviour of people as a supreme value emphasises the rule of the principle of utility, where decisions are made solely depending on whether they have positive or negative consequences while striving to achieve the highest possible benefit for most people (but not necessarily just people).

This principle of utility is usually defined in terms of the amount of suffering and/or enjoyment or happiness (29), meaning that individuals are interested in doing what increases their enjoyment or reduces their suffering. It follows that all living beings, human and non-human, have interests (30). Because all interests, according to this theory, are viewed from a moral viewpoint and deserve equal value, the impact of one's actions on all sensory beings, including animals, is a matter of moral importance. In other words, if someone suffers, it cannot be morally justified to refuse to take that suffering into account (31). As Henry S. Salt, one of the first to advocate for certain animal rights, states: "pain is pain ... whether inflicted on man or beast. And the creature that suffers, whether man or beast, feeling the pain as it lasts, suffers evil" (32). Hence, regardless of the creature's nature, the principle of equality requires that one's suffering be counted as much as the suffering of all other creatures.

For utilitarians, the interests of the highest weight should prevail no matter whose interests they are, and it is precisely this view that has radical measures to assess the greater use of animals (33). Namely, a small step towards a more significant consideration of the interests of animals is better than none. Therefore, according to the utilitarian position, if there are different strategies to improve production, the one that is the best, the most effective will be accepted. In the debate between those who compromise on improving animal welfare and those who seek radical reform, utilitarians do not act on the principle of discussion but consider which strategy will have the best effect on animal welfare. In this context, Signer goes most radical in animal welfare by advocating a boycott of animal products and the settlement of farms by vegetarians. However, this is not because it is fundamentally wrong to kill an animal, but because our consumption of meat and other products from commercially bred animals leads to suffering: "As long as the conscious being is conscious, it has an interest in as much enjoyment and less pain as possible. The feeling is enough for the creature to be brought into the realm of equal consideration of interest. But this does not mean that the creature has a personal interest in continuing to live. For the being who is not self-conscious, death is the cessation of experience, just as birth is the beginning of the experience. Death cannot be contrary to the preference for the continuation of life, while birth can be under the preference for the beginning of life. (...) Since the animal belongs to a species that is not capable of self-awareness, it follows that it is not wrong to breed and kill. The condition is to live a comfortable life, and, after the killing, another domestic animal that will lead a similar life and which would not exist if the first animal was not killed. Vegetarianism is not obligatory for people who eat meat from animals raised in a utilitarian moral way. (...) The essence of utilitarianism is not that it allows killing because it does not belong to the human species, but that it allows killing animals precisely because they lack the ability to want to prolong life. This attitude also applies to members of our species who also do not have that ability" (34).

- **the animal right view**, according to which animals deserve a specific approach that includes the question of what is of best interest to them, regardless of whether people consider them "cute", whether they are helpful to humans, whether they belong to an endangered species or whether at all a person takes care of them (just as a person has his rights even when he is not beautiful, helpful, and even if no one loves him). Philosopher Tom Regan argues that (at least some) animals have negative rights such as the right not to interfere, the right not to be killed, injured or tortured (35), that animals have the right to be treated with respect, then the right to bodily integrity and the right to freedom of movement. Violation of these rights is not morally justified, regardless of the potential benefits people feel they have.

Namely, this approach is based on attributing intrinsic value to all beings who can feel- those who experience life and whose lives can be good or bad over time. As such, they have "individual experiential well-being, logically independent of their usefulness in relation to the interests or well-being of others" (36). Then this is the foundation of their rights and morally obliges us to refrain from things that would significantly hinder the life of such creatures. According to them, the main characteristic that all people have in common is not rationality, but the fact that each of us has his own life that he cares about: what happens to us is important to us, whether it is the same for any who else. This is because we are all subjects of life with experience. Suppose this is really the basis for attributing an inherent value to individuals to be consistent. In that case, we must ascribe an inherent value, and thus a moral right, to all subjects of life, whether human or not.

It follows that an animal rights-based approach is most focused on ensuring animal welfare (experience of pleasure and pain), and attributing protected rights is the best way to achieve this common goal (37). It means understanding that animals are not our property, property that we can use for food, clothing, entertainment or experimentation. Consequently, it is considered wrong to look at animals as a commodity as a "means to an end", just as it is wrong to treat them in the same way for the same reasons. The fundamental right of all who possess an inherent value is the right never to be treated simply as a means to an end for others. In this context, the movement for the protection of animal rights has the same weight as the movement for the protection of human rights.

- **theory of contractarianism and common agreement**, according to which, analogous to the theory of a common agreement of Thomas Hobbes from the 17th century, which claimed that without political rule, everyone would live in a natural state in which our lives would be endangered, the same can be transferred to the use of animals. Proponents of this view claim that because man can establish an "agreement" with other rational beings, that is, with other human beings (because both parties have some benefit from it), and thus protect its rights and interests, with animals he cannot do the same because of their lack of ability to think and make decisions, so it makes no sense to protect their rights because humans get nothing in return.

This is the basis for an argument drawn by analogy that places speciesism side by side with racism and sexism because people as human beings, as a species, hold the view that they are the only ones who deserve moral status or that they at least deserve special moral status is contrary to other species, but without any special justification which is substantiated except that it belongs to the human species (38) and that it is wrong! Namely, the morality of the members of the concluded agreement is applied only to individuals who can agree with the moral community, so it is important to define who those members are. In terms of the moral agreement, morality is a kind of agreement between rational, independent, egocentric individuals who benefit from entering into such an agreement. A vital feature of this view of morality is explaining why we use it and who participates in it. That is, we have it for long-term personal interest, and the parties to the contract in morality include all those who have the following two characteristics:

1. be able to reap certain benefits if they are included in the contract, at least in the long term, if they do not do what they have agreed upon;

2. to be "able" to enter into the contract (39).

Given these requirements, even more, and considering several other alternative attributes such as language abilities, language or speech, rationality, rationality and reasoning, the ability to accept social and moral rules, possession of the immortality of the soul, possession of life in the biographical sense of the word, moral autonomy, the ability for reciprocity, empathy and desire for self-esteem (40) ..., as alleged features that distinguish us from animals and justify our special moral status, it is apparent why animals have no right!

The non-existence of animals in the moral community does not necessarily mean that the way they are treated is irrelevant from the position of a common agreement. Moreover, the position of the common agreement is entirely anthropocentric because any animal right to their protection depends on the human factor (41). From a self-centred point of view, man must necessarily treat animals well enough to suit their needs. Animal suffering is not an ethical problem in itself. Hence, any form of animal use is ethically acceptable and even ethically desirable because of the benefits that humans derive from animals (42).

Conclusion - For a possible bioethics of animals

According to the purpose of this chapter, to understand the central (bio) ethical approaches or perspectives regarding (bio)ethics of animals, i.e. through the use of techniques of experiential and holistic learning to establish an ethical relationship with animals, and hence to be able to recognise and respect our similarities and interrelationships with animals to finally be able to analyse and evaluate the main arguments and directions of thinking that are at the core of (bio)animal ethics – something is more than evident. And that is a prospective treatment of this problem. However, even more so, the emergence of usable solutions for directing social practice requires an experimentally combined presentation within which complements the natural-biocentric and utilitarian point of view, as well as the traditional view of Rousseau and Schopenhauer as the main point of view for the moral conduct, but also the view of the "awe of nature" of Albert Schweizer and Paul Taylor, as additional motivational support. That is, the idea is to find an appropriate solution and apply the two major approaches, i.e., negative utilitarianism and natural biocentrism, a hybrid theory as a new (bio)animal ethics in which it is important to think about both human and animal rights, a theory according to which decisions will be based between respect for nature and animal welfare.

With one acceptance and application of effective measures for action, i.e. those who are in the function of protecting the fundamental animal rights, and also with their legislative operationalisation (43), can prevent further suffering, the pain of the animals, so that their extinction does not occur, which requires the establishment of a benevolent and caring attitude towards them and less extreme forms of their use for human purposes. At the same time, a new culture of human coexistence with other non-human beings should be inaugurated, in accordance with the current living conditions on this planet, i.e. the real threat to biodiversity and the environmental challenges that are here in this first half of the 21st century. The current ecological crisis is simply forcing humanity in a new way to determine its attitude towards animals. And while this may seem utopian, time will tell if humans are ready for this step in evolution, i.e. the first has already been made with the eradication of cannibalism. The second is insight: "will man take the second step by stopping eating animals, i.e. will recognise the fundamental right to life of animals? While this is unlikely to happen in the foreseeable future, this does not mean that man should not work for the recognition of the dignity and protection of non-human living beings" (44). This is because although "modern man in the general humanisation of many spheres of life has significantly surpassed his ancestors, it is still paradoxical that at the same time in our epoch, as in any epoch of mankind so far, the number of animals over which suffering has not been he was so big" (45).

The previous is a result of the fact that "a relevant purposeful socio-economic policy that strives for temporary efficiency requires the inclusion of social considerations and environmental adjustments and incorporating and adjusting to the needs of the animal population and its well-being. Even as part of such an extended perspective and a synthesised view of all the previously enumerated fundamental problematic aspects as a value framework for reforming "real capitalism", we will find ourselves on the path of using a socio-economic policy that will be in line with the needs of living in the 21st century" (46).

References

1. Jakovljević D. „Prava za životinje", *Filozofska istraživanja*, 129 God. 33, Sv. 1 (167–182). Zagreb: HFD, 2003, str. 167-168.
2. Singer P. *In Defense of Animals: The Second Wave*. Malden: Blackwell Publishing, 2006, p. 2.
3. Regan T. *The Case for Animal Rights*. Berkeley: University of California Press, 2004; Regan T. *All That Dwell Therein*. Berkeley: University of California Press, 1982.
4. Meyer-Abich K.M. *Praktische Naturphilosophie*. München: C. H. Beck, 1997; Meyer-Abich K. M. *Wege zum Frieden mit der Natur*. München und Wien: Hanser, 1984.
5. "Universal Declaration of Animal Rights". Available at: http://www.esdaw.eu/ unesco.html.
6. About the concept of "Animal ethics" see: Callicott J. B. and Frodeman R. (eds.) *Encyclopedia of Environmental Ethics and Philosophy*. Farmington Hills, MI: Macmillan Reference USA, 2009, pp. 42-53. Also: Jamieson, D. *Ethics and Environment*. Cambridge: Cambridge University Press, 2008, pp. 112-120.
7. Kaluđerović Ž. "Bioetički pristup životinjama", *Socijalna ekologija*, Zagreb, Vol. 18 (2009.), No. 3-4, str. 312.
8. "Universal Declaration of human rights". Available at: https://civil.org.mk/wp-content/uploads/2013/05/ djeklarazia.pdf.
9. Kaluđerović Ž. "The Reception of the Non-Human Living Beings in Philosophical and Practical Approaches", *Epistēmēs Metron Logos*, Issue 4, No 4, Athens, 2020, pp. 18-31. Also see: Caspar J. *Tierschutz im Recht der modernen Industriegesellschaft*. Baden-Baden: Nomos Verlaggesellschaft, 1999, s. 154.

10. des Jardin J. R. states critical views on Singer's and Regan's views. de Žarden Dž. R. Ekološka etika. Beograd: Službeni glasnik, 2006, str. 193-200. Consult also: McMahan J. *The Ethics of Killing*. Oxford: Oxford University Press, 2002, pp. 194-203.

11. Singer P. *Praktična etika*. Zagreb: KruZak, 2003, str. 205.

12. Aquinas T. *Summa Theologica*, The Second Part of the Second Part, Question 64, article 1. Available at: http://www.newadvent.org/summa/.

13. Sorabji R. *Animal Minds and Human Morals: The Origins of Western Debate*. Ithaca, New York: Cornell University Press, 1995, p. 8.

14. Steiner G. *Anthropocentrism and Its Discontents: The Moral Status of Animals in the History of Western Philosophy*. Pittsburgh: University of Pittsburgh Press, 2005, p. 57.

15. Aristotel, *Nikomahova etika*. Zagreb: Hrvatska sveučilišna naklada, 1992, 1161 a 34 – 1161 b 6.

16. Nussbaum M.C. "Animal Rights: The Need for a Theoretical Basis", *The Harvard Law Review* 114, 5, 2001, pp. 1518-1519.

17. Hume D. *Enquiry Concerning the Principles of Morals*. New York: Oxford University Press, 1975, p. 186.

18. Schopenhauer A. *On the Basis of Morality*. Providence/Oxford: Berghahn Books, 1995, f. 175, 177, 180.

19. As a supplement, futher see Baranzke H. "Tierethik, Tiernatur und Moralanthropologie im Kontext von § 17, Tugendlehre", *Kant-Studien*, Vol. 96, Iss. 3(2005): 336-363.

20. Singer P. *Animal Liberation*, 2nd ed. New York: New York Review of Books, 1990, p. 20.

21. Singer P. *Animal Liberation*. New York: Open Road Integrated Media, Inc., 2009, p. 285 and p. 351. Available at: https://grupojovenfl.files.wordpress.com/2019/10/peter-singer-animal-liberation-1.pdf.

22. Cavalieri P. *The Animal Question: Why Non-Human Animals Deserve Human Rights*. New York: Oxford University Press, 2007.

23. Cavalieri P. "Pushing Things Forward", in: *The Death of the Animal: A Dialogue*, Cavalieri P. et al. (eds.). New York: Columbia University Press, 2009, p. 98.

24. Audi R. *Cambridge Dictionary of Philosophy*. Cambridge: Cambridge University Press, 1999, p. 590.

25. Warren M. A. *Moral Status. Obligations to Persons and Other Living Things*. Oxford: Clarendon Press, 1997, p. 3.

26. De Grazia D. *Prava životinja*. Sarajevo: TKD Šahinpašić, 2004, str. 13.

27. Warren M. A. "Moral status", in: *A Companion to Applied Ethics*, Frey R. G. and Wellman C. H. (eds.). Oxford: Blackwell Publishing Ltd., 2003, p. 439.

28. Vance R. P. "An Introduction to the Philosophical Presuppositions of the Animal Liberation/Right Movement", *Journal of the American Medical Association, JAMA*. 1992; 268(13):1715-1719.

29. Singer P. "All animals are equal", in: *Animal Rights and Human Obligations*, Regan T. & Singer P. (eds.). New Jersey: Prentice-Hall, 1989, p. 150.
30. Engel Jr. M and Jenni K. *The Philosophy of Animal Rights*. Brooklyn: Lantern Books, 2010, p. 14.
31. Singer P. *Practical Ethics*. Cambridge: Cambridge University Press, 2011, p. 50.
32. Salt H. *Animals' Rights: Considered in Relation to Social Progress*. London: George Bell & Sons, 1892, p. 24.
33. Sandøe P. and Christiansen S. B. *Ethics of Animal Use*. Copenhagen: Blackwell Publishing, 2008, p.15.
34. Singer P. "Killing humans and killing animals", *Inquiry*, 22 (1-2), 1979, pp. 145-156.
35. Regan T. *The Case for Animal Rights*. Berkeley: University of California Press, 1983, p. 328.
36. Regan T. "Ill-Gotten Gains", in: *Animal Experimentation: The Consensus Changes*, Langley G. (ed.). London: Macmillan Press, 1989, p. 38.
37. Francione G. L. *Animals as Persons: Essays on the Abolition of Animal Exploitation*. New York: Columbia University Press, 2009, p. 23. Also see Francione G. L. The Animal Rights Debate: Abolition or Regulation. New York: Columbia University Press, 2010.
38. Diamond C. *The Realistic Spirit. Wittgenstein, Philosophy, and the Mind*. Cambridge: MIT Press, 1991, p. 319.
39. Narveson J. "A Case Against Animal Rights", in: *Advances in animal welfare science*, Fox W. M. & Mickley D. L. (eds.). Washington, DC: The Humane Society of the United States, 1986/87, p. 194.
40. Engel Jr. M. and Jenni K. *The Philosophy of Animal Rights*. Brooklyn: Lantern Books, 2010, p. 19.
41. Sandøe P. & Christiansen S. B. *Ethics of Animal Use*. Copenhagen: Blackwell Publishing, 2008, p. 34.
42. Ibid.
43. Take a closer look at Jakovljević D. „Prava za životinje", *Filozofska istraživanja* 129 God. 33 (2013) Sv. 1 (167–182), str. 170-171.
44. Kaluđerović Ž. „Bioetički pristup životinjama", *Socijalna ekologija*, Zagreb, Vol. 18 (2009.), No. 3-4, str. 320.
45. Op. cit., p. 311.
46. Jakovljević D. „Prava za životinje", *Filozofska istraživanja* 129 God. 33 (2013) Sv. 1 (167–182), str. 180-181.

TWO MAIN ETHICAL/MORAL REASONS WHY WE NEED LABORATORY ANIMAL WELFARE –ANIMALS AND HUMANS AS SENTIENT AND CONSCIOUS BEINGS

Marijana Vučinić and Katarina Nenadović

University of Belgrade, Faculty of Veterinary Medicine, Department of Animal Hygiene - Animal Behavior and Welfare, Belgrade, Serbia

Abstract

The welfare of laboratory animals should be ensured for moral/ethical reasons, for the quality of scientific results and for the quality and safety of biomedical products obtained from their use, for the welfare of persons caring for them and using them in researches or testing, but also for public demands and pressure, reputation of institutions that use laboratory animals, etc. Two sentient and conscious beings - animals and humans - participate in research and testing on laboratory animals. Their well-being is common because only from animals that are spared the inconveniences that change their neuroendocrine, immune, metabolic status and behavior, quality biomedical results and products can be obtained.

Keywords: laboratory animals, welfare, ethical/moral reasons

Introduction

Probably, by ceasing to use nonhuman animals (animals), we would free them from all the sources of unpleasant experiences we bring them into (1). However, because we still use them, we must provide all the conditions to prevent their unnecessary suffering. Therefore, it is our obligation to provide all the mechanisms for their welfare during their life and at the time of death. Killing animals is an ethical issue, starting with questions why we kill animals and how we kill them. The way animals die is also a matter of their welfare (2, 3). We do not want products from suffering animals, nor do we want scientific or testing results obtained from laboratory animals that suffer. We are not able to establish a bidirectional positive emotional connection with animals that suffer. However, we need animal welfare not only for animals but also for us as animal users, our own well-being, safety and security, our satisfaction, quality of products or results obtained by animal using, our reputation, economy, sustainability, public demands and many other reasons. In this chapter, we will try to explain why we need the welfare of laboratory animals, starting with the way people use animals, through the procedures that conduct on laboratory animals.

The ways we use animals

Human beings use a large number of non-human animals (animals) for own benefits in different ways (4, 5). They use animals for food production, the production of natural fibers, transport, for physical work, as service animals, for entertainment, as pets and companions, in education, basic and translational science and testing the effectiveness, harmfulness and safety of various products, but also for treatment of their own health disorders including both, physical and mental conditions (6).

People take care of the animals they use by providing them with accommodation and shelter from adverse weather conditions, predators and other sources of various dangers, a place for resting and sleeping, food, water, treatment or prevention of health disorders and diseases and other necessities for survival. However, humans also intentionally harm animals disrupting their physical, mental and emotional integrity (7). To get food, some animals that people take care of are slaughtered. People who care for animals have to catch and lift to examine, treat or mark them. Some animals are neutered to prevent them from giving birth to unwanted offspring. Some animals, mostly pets that have not met the expectations of their owners, people abandon illegally, leaving them to all the dangers lurking in public places. In addition, people take care of many animals in an enclosed environment that they have designed and equipped according to their own needs and the needs of the animals. Very often, in this half-half designed environment animals are exposed to various sources of unpleasant experiences, discomfort and insecurity that come from various phenomena, objects, materials, other animals or people. In addition, people separate the cubs from their mothers, put some animals in social isolation, and prepare meals for some that do not create a feeling of satiety and physical, physiological or emotional satisfaction in animals. Some animals have to be completely deprived of food due to their own safety and safety of people who will use animal-derived products (for example, before and during transport or before slaughter).

Enclosed in an environment designed by the people who care for them, animals are often deprived of many other stimuli, space or circumstances that motivate them to physical activity and to manifest natural behavior. This means that some of animals are prevented from anticipating and controlling events in their living environment and from independently satisfying their needs by choosing the necessary substrate and the most effective behavioral strategy. Therefore, certain actions of people towards the animals they take care of and certain events in the living environment of animals can cause frustrations, fears, distress, pain, suffering, boredom and many other inconveniences and unpleasant physical or emotional experiences in them. On the other hand, there are new ways of unintentional indirect harms that people cause to animals injuring and killing vast numbers of them by disturbing the balances and processes of nature through pollution, introduction of alien/invasive species and climate change or by human artifacts such as cars, windows and communication towers (7, 8). Therefore, it seems that the 21st century and high technologies that people use need new responsibilities toward animals, new ways of interactions with animals (5) and a new ethics for animal use (8). Simply put, we need changes in the direction of a more humane and rational society.

People keep the animals they take care of in their homes, on farms, in kennels, in shelters, boarding houses, zoos or laboratories, but also in many other places. The care and use of animals inevitably leads to a certain interaction and relationship between humans and the animals they care for (9). Humans always benefit more from this interaction than animals because humans use them for specific purposes. If they could choose, animals would certainly not choose to be used in the ways that humans use them. This is especially true for laboratory animals, to which people who use them cause a variety of physical and emotional inconveniences in the name of basic and translational biomedical science striving to study how bodies of animals and humans function; to develop and test new medicines and vaccines for humans or animals or new diagnostic methods or treatments of diseases and to assess the efficacy, harmfulness and safety of various products for human and animal health and environment (10, 11).

What we do to laboratory animals?

Personnel working in laboratories take care of laboratory animals. They keep laboratory animals that wait to be included in experiments in safe living conditions feeding and watering them, protecting them from injures and diseases and taking care of the ambient and hygienic conditions in animal living space. They prepare laboratory animals for residence in the institutional animal care facilities, and for the procedures that the animals will undergo in the experiments in order to minimize unpleasant experiences in lab animals during experiments. Preparing animals for unpleasant procedures also means behavioral conditioning including their habituation to, desensitization to, and training for procedures that will be involved in experiments (10). Everything we do with animals that are not yet directly included in the experiment, we do to avoid all the inconveniences that can unforeseen negatively affect their physical and mental welfare and the results of the experiment. The precondition is that we have chosen the appropriate animal model and applied the "3Rs rule"- Replacement, Reduction, Refinement (11). However, the question now is what we do to animals to make appropriate models out of them and what we do with animals in experiments and testing?

Laboratory animals spend their entire life in the laboratory, housed in cages or boxes. They are provided with food, water, bedding material, and a place to rest and sleep. In order to avoid the development of pathological forms of behavior caused by boredom and very limited physical and mental activity, a strategy of enriching living conditions is applied by inserting stimuli (objects, materials, phenomena) that motivate them to be physically and behaviorally active. This means that laboratory animals confined in strictly controlled living conditions are given the opportunity to manifest natural forms of behavior, including forms of behavior highly specific to the species (12).

Laboratory workers catch laboratory animals, lift them, restrain them, move them to another accommodation units, change their bedding, clean cages and boxes, regroup them, socially isolate them, mark them, and sample their body fluids and tissues, anesthetize and euthanize them.

In order to make appropriate animal models and achieve appropriate results in biomedicine, we change their phenotypic and genotypic characteristics. We modify them genetically and change their microbiological composition. We also change their immune and metabolic profile and structure of their tissues and organs. We harm laboratory animals to study the mechanisms of recovery from injuries and to study drugs and therapies that promote and accelerate the healing of wounds. By an artificial way, we cause non-infectious and infectious diseases in them in order to study the mechanisms of development and consequences of diseases, and to find medicines and biological means against those diseases. On laboratory animals we test toxic chemicals, devices and biological substances that may cause severe patho-morphological and patho-physiological changes in their organs with severe pain, distress or death. We produce tumors in them allowing tumors to cause cachexia, to spread the rough a body, or to ulcerate. There are many others examples of procedures that laboratory workers and researchers conduct on laboratory animals accompanied with severe pain, distress and death. In order to prevent suffering in laboratory animals those who use them should know legislatives, how to plan procedures that will conduct on laboratory animals, severity classification, severity assessment of procedures and the harm–benefit analysis of projects, how to develop and implement a system for the monitoring and assessment of laboratory animal welfare, how to develop and implement a suitable recording system and when to stop procedures on animals in order to prevent suffering in them (13). Otherwise, there may be severe suffering of animals, lack of expected results and disappointment of workers and researchers related to feelings of guilt, remorse, regret, grief, and loss of self-confidence, general fear of failure, fear of losing a job, etc (14). All this is the reason why it is necessary to apply mechanisms for ensuring the welfare of laboratory animals.

What is animal welfare?

Welfare is a condition that shows how an animal cope with the living conditions provided for it by the person who takes care of it (15). This condition can be assessed on the animal-based indicators such as the appearance, behavior and feelings of the animals (16, 17). It can be also assessed on the resource-based indicators such as living conditions, i.e. on the basis of housing conditions, including space, microclimate, objects, lighting, smells, and noises, materials, substrates and, hygienic status provided by persons who take care for animals or naturally appear in an animal living surrounding. Ensuring welfare is achieved by providing an animal with good nutrition, good housing, good health and appropriate behavior, including physiological behaviors, the absence of behavioral disorders and pathological behaviors, and behavior toward other animals and humans. Animals cannot adapt to bad living condition, and they are a constant source of negative states, feelings and experiences (16). Such are fear and states similar to fear, anxiety, stress and distress, frustrations, conflicts, deprivations, pain, physical and thermal discomfort, insecurity, dissatisfaction, boredom, suffering, thirst, hunger, weakness, debility, breathlessness, nausea, sickness, helplessness, loneliness etc (17). When people decide to use animals for production, testing, researches or for company purposes, then they decide to put them in certain living conditions that they think suit the animal. Negative experiences in animal are an inevitable occurrence in the living conditions imposed on animals, and if we cannot avoid them, we must try to keep them to a minimum. Therefore, animals cannot adapt to poor living conditions. They will cope with them until they become physically, physiologically and emotionally exhausted, which will result in a total lack of welfare. The consequence of that will be an emaciated animal with a bad body condition, through diseases of non-infectious and infectious etiology and disorders and the emotionally exhausted suffering animal with pathological forms of behavior.

The goal of providing welfare to animals is to protect an animal integrity including genetic, physical, physiological and mental integrity. It can be achieved by minimizing negative activities by human beings and states, feelings, and experiences related to life in an ambient designed by humans and by achieving positive states, feelings, and experiences such as states of comfort, pleasure, security, and physical, physiological, and emotional satisfaction. Also, one should avoid any practice that violates any integrity of the animal. However, is it possible in circumstances in which human beings use animals with a certain goal and intention?

There are three groups of factors affecting animal welfare. The first of them comes from the animal itself including an animal genetics. The second includes environmental factors in the animal surrounding or its living conditions and finally, the third group embraces human-animal relationship (18). Considering the fact that the owner chooses the animal based on its characteristics such as appearance and behavior, productive or reproductive characteristics, as well as that the owner determines the living conditions of the animal, it can be said that the owner affecting all factors that impact animal welfare including genetics. When the genetic integrity of laboratory animals is intentionally damaged by genetic manipulations, then it is in order to obtain better animal models for the study of human diseases and the provision of new medicines and therapies (19). If a quality model is obtained, then such a model significantly reduces the use of laboratory animals of other species and strains, which satisfies the requirement to reduce the number of animals according to the 3Rs rule. It is very important to ensure the welfare of the animals in the experiment. However, a large number of animals have been housed in experimental facilities and are awaiting inclusion or have some other roles in the experimental institutions. Living conditions in unstimulating environments of cages and boxes are mostly boring for animals. It is not uncommon for them to develop stereotypic forms of behavior (20, 21). It was documented that abnormal behavior has potential to affect experimental outcomes compromising their validity, reliability, and replicability, especially in behavioral experiments (20). Therefore, despite animals directly involved in an experiment, it is important to take into account the welfare of animals that are not directly involved in the experiment (22). These animals are from breeding stocks and breeding reserves or present biological surplus, managed surplus, sentinel, accompanying, waiting, and post-experimental groups. Just as it is important to improve and refine all the procedures that will be performed on the animals in the experiment, so it is important to improve the living conditions of animals that are not in the experiment and live in the same institution as the animals in the experiment. From this point of the view, different types of enrichment may be used such as structural, social, foraging and feeding, manipulating enrichment and many others. Enriched environments have positive effect on the welfare of experimental animals including behavioral, physiological and neurochemical characteristics. Recent research has found that mice housed in conventional conditions showed more inactive and stereotypical behavior than mice housed in enriched conditions (23). It is well known that inappropriate living condition of experimental animals may deteriorate the quality of results obtained in scientific studies (22). Moreover, variation in stereotypic behaviour may represent an important source of variation in many animal experiments (21). This is because different species and strains of experimental animals can develop different forms of stereotypes. Not all forms of stereotypes reflect negative affective states and negative cognitive bias. It depends on experimental animal strains.

Human-animal relationship is it also very important for the laboratory animal welfare. It is well known that handling technique and handling frequency can be an important source of stress and anxiety in laboratory animals. Thus, for mice, tunnel handling is less stressful than tail handling technique (24, 25). Gentle handling can foster a better relationship between the handlers and rodents and reduce depressive-like behavior in mice (26). This is of particular importance both for the welfare of laboratory animals and for the well-being of people caring for and using laboratory animals (27).

The role of behavior is to facilitate homeostasis by rapid adaptation of an animal to changes in its environment or allowing the animal to control and modify its environment (20). However, in a captive environment behavioral flexibility may be reduced and animals may be unable to cope with changes. Moreover, they will try unsuccessfully to cope by exhausting their adaptation mechanisms, which will frustrate them, bring them into a state of distress and change their physiological, emotional and behavioral profile. In such cases, previously flexible behavior can take the form of abnormal behavior. Abnormal behavior may be associated with abnormal physiology affecting experimental outcomes.

It is certain that experimenting with animals impairs their welfare. Such impaired welfare can turn into animal suffering. If our humanity and technology have not evolved towards the use of non-animal alternatives and we still use animals, then it is our moral obligation and responsibility to reduce or avoid any animal suffering. We have to know what, when and how to do it in order to prevent suffering in laboratory animals. In other words, just as important as it is to be responsible, it is just as important to be well educated and skilled regarding the handling of laboratory animals.

Humans need the welfare of laboratory animals because they are conscious and sentient beings

There are many reasons why we need the welfare of laboratory animals. Firstly, laboratory animals are conscious and sentient beings. To be conscious means to have subjective experiences about the surrounding world and own body (28, 29). Conscience serves animals to orient themselves and adapt to their living environment. Laboratory animals, like humans, have the capacity to feel unpleasant and pleasant experiences, to memory them and to decide how to react in the next encounter with same stimuli that caused discomfort or pleasure or in the circumstances in which they experienced these feelings. It is very important for them to live in conditions in which they can anticipate events and control their living environment. However, experimental animals are often used in a way and kept under living conditions where their power of prediction and control is not possible or is minimized. This undermines their security and makes them feel uncomfortable and unsecure. It can be said that unpleasant procedures with laboratory animals change their emotions and their consciousness. The cognitive bias test reflects the current emotional state of the individual and may be a suitable method for severity assessment in laboratory animals investigating the influence of their previous experiences on the expectation of future events. It means that laboratory animals which have experienced negative events expect additional negative events and react more hesitantly towards new situations (30).

Experimental animals have their right to life. If we cannot replace them with non-animal alternatives, then we must ensure that they live well. By deciding to use them in scientific experiments and testing, we decided to influence their lives and expose them to a series of inconveniences that they would never have chosen. That is why we must reduce these inconveniences to a minimum. We use different species of animals in research and testing. It is not possible to grade consciousness in relation to the kind of animal because each species has its own distinctive consciousness profile and interspecies variations are possible due to differences in perceptual richness, evaluative richness, integration at a time, and integration across time, and self-consciousness (29). We believe that all vertebrates used as laboratory animals possess some dimensions of consciousness and self-awareness.

Recently, this was supported by the mirror test that the fish (*Labroides dimidiatus*) recognized that their reflections in the mirror belonged to their own bodies and checked themselves in the mirror before and after the scraping (31).

The role of consciousness is to adapt (32), shape and plan the behavior of an animal-based on the animal's own decision in real space and time depending on real events and situations in its living environment in which there are various phenomena, objects, materials, and beings. This behavior must be beneficial to the animal in relation to the events in its environment. Consciousness has its key features and indicators (28). Key features that characterize consciousness are qualitative richness, situatedness, intentionality, integration, dynamics and stability. These features are necessary elements for consciousness. On the other side are indicators of consciousness. They can be found in brain anatomy and physiology, goal directed behavior and model-based learning, psychometrics and meta-cognitive judgment, episodic memory, illusion and multistable perception and visuospatial behavior (28).

This is why animals are sentient beings with the capacity to feel unpleasant physical and emotional experiences, the interests of animals must therefore be taken into consideration. This means that people who use laboratory animals should spare and protect them from all inconveniences before and during the experiment or testing itself as well as after these events.

Some scientists avoid using terms such as emotions, sentiments, feelings and moods in animals, believing that it is anthropomorphism. Instead of the above terms, they use the term "affective states". This term refers to the condition of the animal individual perceptions of internal and external stimuli. Affective states can be both positive and negative. Affective states motivate animals to certain behaviors, and experienced caretakers and scientists who work with laboratory animals can interpret some of them precisely on the basis of behavior, physiologic parameters, body language and facial expression of the animal. Some of affective states in animals are not possible to assess (17). Negative affective states are contrary to animal welfare, disrupting it (33, 34). Particularly negative affective states in animals are pain, distress, thirst, hunger, nausea, breathlessness, dizziness, debility, weakness, sickness, anxiety, boredom, fear, frustration, helplessness, and loneliness (17). To this list can be added discomfort, inconvenience, insecurity and dissatisfaction. Some of positive emotional states in animals are satiety, vitality, reward, contentment, curiosity, and playfulness (17) but also experiences of comfort, pleasure, security and satisfaction. Many procedures with laboratory animals are unpleasant, stressful, painful and frightening for them. The way laboratory animals are housed can cause feelings of discomfort and insecurity. Life in captivity is often associated with frequent exposure to stressors, which may be the source of persistent negative affective states in laboratory animals causing captivity-induced depression (35) or abnormal behaviours (20). Behavior change is the fastest way to adapt to changes within the organism and in the living environment of animals. Any change threatens to disrupt the homeostasis of the organism and that is why the role of behavior is to preserve the homeostasis of the animal's organism. By changing behavior, animals control and modify conditions in their living environment. Laboratory animals can never adapt to some living conditions and the procedures that are carried out on them. Therefore, they will be constantly under the influence of factors that threaten to disrupt the homeostasis of their organism (stressors), which will result in a change in their behavior, physiology and their emotional experiences. Abnormal behavior is an indication that laboratory animals cannot adapt to living conditions that are rich in aversive stimuli, including housing conditions and the procedures performed on them. Stereotypies and other abnormal repetitive behaviors are common in laboratory animals. These abnormal behaviors have a potential to affect validity, reliability, and replicability of scientific outcomes (20).

Why laboratory animal personnel need animal welfare?

There are at least two reasons why we need the welfare of laboratory animals when it comes to the people who work with them. The first refers to the physical health, and the second to the mental health of workers with laboratory animals. It was experimentally confirmed that that participation in animal experiments can negatively impact the mental health of researchers increasing anxiety scores in animal users comparing with non-animal users in both genders, male and female. Also, younger animal users with less work experience and lower income level exhibited higher anxiety scores than non-animal users (36, 37). It was found that young people involved in animal testing, feel remorse, emotional tension and helplessness (37).

People who work with laboratory animals are exposed to various health hazards and risks that arise from the animals themselves, equipment, accessories, and materials they use, as well as the technical and construction characteristics of the rooms in which they work and through which they move. These hazards and risks include animal-related, sharp-related, noise-related, radiation-related, and electricity-related injuries, injuries from slipping, tripping and falls, explosions and fires, poisonings with toxic substances, allergies, burns and infections and zoonoses (38, 39, 40). Persons who handle laboratory animals in any way need animal welfare. They need laboratory animal welfare for several reasons. First, as conscious and sentient beings and remembering beings, laboratory animals will try to avoid any inconvenience they have remembered in previous treatment with them. If they can't escape the inconvenience caused by the people who handle them, they will try to protect and defend themselves. In that defense, they can bite or scratch the people who handle them. Therefore, at this moment, there is a danger of transmitting zoonoses from animals to humans (41). Working with laboratory animals does not present only a risk of injuries and zoonoses (42), but also a risk of toxic substances and allergies (Nicholson). Younger and less experienced workers are particularly exposed to these risks (38).

It is a generally accepted fact that the workplace can bring a pleasure and satisfaction, but it can also be a source of many risks, inconveniences, fatigue, stress and general exhaustion for workers. Laboratory animal caretakers, technicians, welfare officers and veterinarians and researches participate in the work with laboratory animals. Their job satisfaction and professional quality of life are influenced by numerous factors including intercollegiate relationships and interactions, but also the relationships and interactions with laboratory animals (43).

Personnel working with laboratory animals are exposed to multifactorial stressors. The impact of stressor can have negative consequences for the mental health of employees, their quality of life, motivation to work and interaction with animals. In these stressors Murray et al. (44) included compassion and moral stress, issues related to staffing and scheduling of work, insufficient communication in the workplace, and public ambivalence toward the use of animals in science. These work stressors can leave caring and compassionate employees feeling alone, anxious, and unsupported. It is well known that the welfare of experimental animals also affects the well-being and quality of life of staff working with them. The study conducted in American and Canadian laboratory animal personnel found that personnel reporting poorer professional quality of life also reported less social support, higher animal stress/pain, less enrichment diversity/frequency and wished they could provide more enrichment, using physical euthanasia, and less control over performing euthanasia (27). Compassion stress, moral stress and euthanasia stress poses a special danger to the physical, emotional and psychological exhaustion so called compassion fatigue of staff working with experimental animals (45, 46). Interesting results were obtained recently by Spanish investigators who study professional quality of life in researchers using laboratory animals (43). They found that animal-facility personnel showed significantly higher total professional quality of life and compassion satisfaction scores than researchers. However, compassion fatigue score were higher in researchers. Other studies also confirmed that most influential work-related factors associated with feelings of compassion fatigue were understaffing, close relationships with experimental animals, a lack of resources for coping with compassion fatigue, poor relationships with superiors, and lack of training in managing compassion fatigue (47).

Emotional burnout and staff exhaustion are mainly due to conflicts of emotions that arise during animal care on the one hand and injuring and sacrificing/killing those animals from which tissue samples need to be taken, or those that cannot be reused or those that cannot be adopted. This conflict of emotions got its name "caring-killing paradox" (48). Particular emotional exhaustion can be developed in persons who care for laboratory animals but not conduct experiments directly. These persons often become emotionally attached to laboratory animals, transforming them into their pets. Emotional attachment to laboratory animals can be a cause of conflicts between technicians and researchers directly involved in animal experimentations (49). Emotional burnout and personnel emotional exhaustion can be linked to poor animal treatment and thus may jeopardize laboratory animal welfare. This is another reason that explains why we need laboratory animal welfare (50, 51) and the concept of "One welfare" (52).

The inconveniences associated with the workplace of personnel using laboratory animals may also be a consequence of the activities of the animal protection movement and the public demands. The pressures and demands of the public in relation to the use of laboratory animals are especially unfavorable for the laboratory animal personnel. Sometimes, their profession is viewed as morally tainted. They may be particularly intimidated by the activism of extreme animal advocates and may be especially afraid of stigmatizing their job by negative public perceptions (53). The public reacts differently to the use of laboratory animals, which mainly depends on the kind of animal and the type of research (54), but on impact on animals, humans and, scientific merit and availability of alternatives (55). A survey conducted in European countries found that experimentation on animals such as mice was generally accepted, but public perceptions were divided on dogs and monkeys (56). Thus, the public justifies the use of animals for testing drugs and vaccines rather than for testing non-essential goods for human life, such as cosmetics (55). The public and animal protectors are interested in openness and transparency of work with laboratory animals. This can be achieved in a variety of ways, and there are suggestions that animal advocates be directly involved in working with laboratory animals giving ideas how to improve the quality of laboratory animal use (57).

If we want to open our laboratories to the public and make our work openness and transparent, especially for animal advocates, the first thing we need to respect is animals and their welfare. In that way, we could wash the stain off our work caused by misunderstanding and ignorance of working conditions with laboratory animals and show the public and animal advocates that we take care on animals in laboratories in the most humane way. It is true that demands for openness and transparency in the work with laboratory animals make users of laboratory animals more responsible. This is justified from the aspect of the welfare of laboratory animals, but it represents a strong pressure on the workers. However, increasing transparency in work with laboratory animals and institutions which use them could result in a more positive perception of lab animal researchers and the work that they do (53). Otherwise, laboratories that use animals can stop working and be closed, lose their reputation, interrupt their sustainability, which calls into question the financial status of workers and scientists. Moreover, it was confirmed recently that openness about animal research even on primates increases public support (58).

Conclusion

We use laboratory animals for various purposes for our own benefit. The damage we cause to laboratory animals is numerous and comes down to intentional and deliberate disruption of their physical, genetic or emotional integrity. Such actions harm the welfare of laboratory animals, and ethical / moral reasons force us to harm their welfare as little as possible. Otherwise, the outcomes of using laboratory animals will be poor and unreliable scientific results or poor quality biomedical products. Such outcomes will inevitably affect the well-being of laboratory animal care staff and researchers using laboratory animals.

References

1. Francione GL. Animal rights theory and utilitarianism: relative normative guidance. Animal L. 1997; 3: 75-101.
2. Yeates, J.W. Death is a welfare issue. J. Agric. Environ. Ethics. 2010; 23: 229–241.
3. Butterworth A, Yeates J. (2018). Longevity and brevity - is death a welfare issue? In A. Butterworth (Ed.), Animal welfare in a changing world (pp. 190). CABI Publishing.
4. Nawroth C, Baciadonna L, Emery NJ. Editorial: Humans in an animal's world - how non-human animals perceive and interact with humans. Front. Psychol. 2021; 12: 733430. doi: 10.3389/fpsyg.2021.733430
5. Tarazona AM, Ceballos MC, Broom DM. Human relationships with domestic and other animals: one health, one welfare, one biology. Animals (Basel). 2019; 10(1): 43. doi:10.3390/ani10010043
6. Wünderlich NV, Mosteller J, Beverland MB, et al. Animals in our lives: an interactive well-being perspective. J. Macromarketing 2021; 41(4):646-662. doi: 10.1177/0276146720984815
7. Fraser, D. A "practical" ethic for animals. J. Agric. Environ. Ethics. 2012; 25(5), 721-746.
8. Fraser D. Why we need a new ethic for animals. J. Appl. Anim. Ethics Res. 2019; 1(1): 7-21. doi: 10.1163/25889567-12340002
9. Rault J-L, Waiblinger S, Boivin X and Hemsworth P. The power of a positive human–animal relationship for animal welfare. Front. Vet. Sci. 2020; 7: 590867. doi: 10.3389/fvets.2020.590867
10. Todorović Z, Prostran M, Medić B et al. Bioethics and pharmacology. In: Bioethics and Pharmacology: Ethics in Preclinical and Clinical Drug Development (Editors: Todorović Zoran, Prostran Milica, Turza Karel), Published by Transworld Research Network, Trivandrum-695 023, Kerala, India, 2012, 7-14.

11. Vučinić M, Trailović S, Todorović Z et al. Ethics of animal use in preclinical phase of drug testing. In: Bioethics and Pharmacology: Ethics in Preclinical and Clinical Drug Development (Editors: Todorović Zoran, Prostran Milica, Turza Karel), Published by Transworld Research Network, Trivandrum-695 023, Kerala, India, 2012, 15-33.

12. Schapiro SJ, Everitt JI. Preparation of animals for use in the laboratory: issues and challenges for the institutional animal care and use committee (IACUC). ILAR J. 2006; 47(4): 370-375.

13. Russell WMS, Burch RL. The Principles of Humane Experimental Technique. 1959; London: Methuen and Co Ltd.

14. Ratuski AS; Weary DM. Environmental enrichment for rats and mice housed in laboratories: a metareview. Animals (Basel) 2022; 12(4): 414. doi: 10.3390/ani12040414.

15. Smith D, Anderson D, Degryse A-D et al. Classification and reporting of severity experienced by animals used in scientific procedures: FELASA/ECLAM/ESLAV Working group report. Lab. Anim. 2018; 52(1 suppl): 5-57.

16. Vučinić M, Lazić I. Animal welfare assessment. Vet. glasnik 2008; 62 (1-2): 97-104.

17. Hawkins P, Morton DB, Burman O et al. A guide to defining and implementing protocols for the welfare assessment of laboratory animals: eleventh report of the BVAAWF/FRAME/RSPCA/UFAW Joint Working Group on Refinement. Lab. Anim. 2011; 45(1): 1-13.

18. Kranke N. How the suffering of nonhuman animals and humans in animal research is interconnected. JAE 2020; 10(1): 41-48.

19. Broom DM. Indicators of poor welfare. Vet. J. 1986; 142: 524-526.

20. Broom DM. Coping, stress and welfare. In: Coping with challenge: welfare in animals including humans. Proceedings of Dahlem Conference, ed. D.M. Broom, Dahlem University Press , Berlin 2001; 1-9.

21. Mellor DJ. Affective states and the assessment of laboratory-induced animal welfare Impacts. Altex Proc. 2012; 1: 445-449.

22. Baumans V. Science-based assessment of animal welfare: laboratory animals. Rev. - Off. Int. Epizoot) 2005; 24(2): 503-513.

23. de Graeff N, Jongsma KR, Johnston J, et al. The ethics of genome editing in non-human animals: a systematic review of reasons reported in the academic literature. Phil. Trans. R. Soc. 2019; B 374 (1772): 20180106. doi:10.1098/rstb.2018.0106

24. Garner JP. Stereotypies and other abnormal repetitive behaviors: potential impact on validity, reliability, and replicability of scientific outcomes. ILAR J. 2005; 46(2): 106-117.

25. Novak J, Bailoo JD, Melotti L, Würbel H. (2016). Effect of cage-induced stereotypies on measures of affective state and recurrent perseveration in CD-1 and C57BL/6 mice. PloS one 2016; 11(5): e0153203. doi:10.1371/journal.pone.0153203

26. Lewejohann L, Schwabe K, Häger C, Jirkof P. Impulse for animal welfare outside the experiment. Lab. Anim. 2020; 54(2):150-158.

27. Hobbiesiefken U, Mieske P, Lewejohann L, Diederich K. Evaluation of different types of enrichment - their usage and effect on home cage behavior in female mice. PloS one 2021;16(12): e0261876. doi:10.1371/journal.pone.0261876

28. Gouveia K, Hurst JL. Improving the practicality of using non-aversive handling methods to reduce background stress and anxiety in laboratory mice. *Sci. Rep. 2019*; **9:** 20305. doi: 10.1038/s41598-019-56860-7

29. Sensini F, Inta D, Palme R et al. The impact of handling technique and handling frequency on laboratory mouse welfare is sex-specific. Sci Rep. 2020; 10(1): 17281. doi: 10.1038/s41598-020-74279-3.

30. Neely C, Lane C, Torres J. et al. The effect of gentle handling on depressive-like behavior in adult male mice: considerations for human and rodent interactions in the laboratory. Behav. Neurol. 2018; 2976014. doi: 10.1155/2018/2976014.

31. LaFollette MR, Riley MC, Cloutier S. et al. Laboratory animal welfare meets human welfare: a cross-sectional study of professional quality of life, including compassion fatigue in laboratory animal personnel. Front. Vet. Sci. 7; 114. doi: 10.3389/fvets.2020.00114

32. Pennartz CMA, Farisco M, Evers K. Indicators and criteria of consciousness in animals and intelligent machines: an inside-out approach. Front. Syst. Neurosci. 2019; 13: 25. doi: 10.3389/fnsys.2019.00025.

33. Birch J, Schnell A, Clayton N. Dimensions of animal consciousness. Trends. Cogn. Sci. 2020; 24(10): 789-801.

34. Kahnau P, Habedank A, Diederich K et al. Behavioral methods for severity assessment. *Animals (Basel)*. 2020; 10(7):1136. doi:10.3390/ani10071136.

35. Kohda M, Hotta T, Takeyama T *et al*. If a fish can pass the mark test, what are the implications for consciousness and self-awareness testing in animals? PLoS Biol. 2019; 17(2): e3000021. doi: 10.1371/journal.pbio.3000021

36. Peper A. A general theory of consciousness I: consciousness and adaptation. Commun. Integr. Biol. 2020; 13(1): 6-21.

37. Jirkof P, Rudeck J, Lewejohann L. Assessing affective state in laboratory rodents to promote animal welfare-what is the progress in applied refinement research?. Animals (Basel). 2019; 9(12):1026. doi:10.3390/ani9121026.

38. Kremer L, Klein Holkenborg S, Reimert I et al. The nuts and bolts of animal emotion. Neurosci. Biobehav. Rev. 2020; *113*: 273–286.

39. Lecorps B, Weary DM, von Keyserlingk M. Captivity-induced depression in animals. Trends Cogn. Sci. 2021; 25(7): 539–541.

40. Kang M, Han A, Kim DE et al. Mental stress from animal experiments: a survey with Korean researchers. Toxicol. Res. 2018; 34(1):75-81

41. Mamzer H, Zok A, Białas P et al. Negative psychological aspects of working with experimental animals in scientific research. PeerJ 2021; 9: e11035. doi: 10.7717/peerj.11035

42. Bibay JIA, Agapito JD. Survey on health and safety concerns of laboratory animal workers in the Philippines. Philipp. J. Sci. 2022; 151(2): 605-614.

43. Dyson MC, Greer WG, Colby LA. Institutional responsibilities for the oversight of personnel safety in animal research. Appl Biosaf 2018; 23(3): 122-129.

44. Nicholson PJ, Mayho GV, Roomes D et al. Health surveillance of workers exposed to laboratory animal allergens. Occupational Medicine 2010; 60(8): 591–597.

45. Weigler BJ, Di Giacomo RF, Alexander S. A national survey of laboratory animal workers concerning occupational risks for zoonotic diseases. Comp. Med. 2005; 55(2): 183-191.

46. Steelman ED, Alexander JL. Laboratory animal workers' attitudes and perceptions concerning occupational risk and injury. J Am Assoc Lab Anim Sci 2016; 55(4): 419-425.

47. Goñi-Balentziaga O, Vila S, Ortega-Saez I et al. Professional quality of life in research involving laboratory animals. Animals (Basel) 2021; 11(9): 2639. doi: 10.3390/ani11092639

48. Murray J, Bauer C, Vilminot N. et al. Strengthening workplace well-being in research animal facilities. Front. Vet. Sci. 2020; 7:573106. doi: 10.3389/fvets.2020.573106

49. Newsome JT, Clemmons EA, Fitzhugh DC et al. Compassion fatigue, euthanasia stress, and their management in laboratory animal research. JAALAS 2019; 58(3): 289–292.

50. Van Hooser JP, Pekow C, Nguyen HM et al. (2021). Caring for the animal caregiver-occupational health, human-animal bond and compassion fatigue. Front. Vet. Sci. 2021; 8: 731003. doi: 10.3389/fvets.2021.731003

51. Randall MS, Moody CM, Turner PV. Mental wellbeing in laboratory animal professionals: a cross-sectional study of compassion fatigue, contributing factors, and coping mechanisms. JAALAS 2021; 60(1): 54-63.

52. Reeve CL, Rogelberg SG, Spitzmüller C, et al. The caring-killing paradox: euthanasia-related strain among animal-shelter workers 1. J. Appl. Soc. Psychol. 2005; 35(1): 119–143.

53. Herzog H. Ethical aspects of relationships between humans and research animals. Ilar J 2002; 43(1):27–32.

54. Vučinić M. Basic principles of experimental animal welfare protection. Vet. glasnik 2007; 61(3-4): 173-181.

55. Vučinić M, Radenković-Damnjanović B, Radisavljević K. What is and why we need animal welfare? Vet.J. RS (Banja Luka) 2011; 11(1): 59-68.

56. Pinillos RG, Appleby MC, Manteca X et al. One welfare - a platform for improving human and animal welfare. Vet. Rec. 2016; 179(16): 412–413.

57. Mills KE, Han Z, Robbins J et al. Institutional transparency improves public perception of lab animal technicians and support for animal research. PLoS ONE 2018; 13(2): e0193262. doi: 10.1371/journal.pone.0193262

58. Ormandy EH, Schuppli CA. Public attitudes toward animal research: a review. Animals (Basel) 2014; 4(3): 391-408. doi: 10.3390/ani4030391.
59. Brunt MW, Weary DM. Public consultation in the evaluation of animal research protocols. PLoS One. 2021; 16(12): e0260114. doi: 10.1371/journal.pone.0260114
60. von Roten FC. Public perceptions of animal experimentation across Europe. Public. Underst. Sci. 2013; 22(6): 691-703.
61. Carbone L. Open transparent communication about animals in laboratories: dialog for multiple voices and multiple audiences. Animals (Basel) 2021; 11(2): 368. doi: 10.3390/ani11020368.
62. Mendez JC, Perry BAL, Heppenstall RJ et al. Openness about animal research increases public support. Nat. Neurosci.2022. doi:10.1038/s41593-022-01039-z

REGULATIONS AND ETHICAL CONSIDERATIONS IN ANIMAL EXPERIMENTS – THE CASE OF EUROPE

Siniša Đurašević

University of Belgrade Faculty of Biology, Belgrade, Serbia

Abstract

The history of European legislation on the use of laboratory animals in some countries dates back to the 19th century. Over time, several documents and laws have been passed to regulate the use of animals in experiments. In 1985 an agreement has been reached in Strasbourg on the Convention on the Protection of Vertebrate Animals Used for Experimental and Other Scientific Purposes (ETS123). With the primary aim of reducing a number of animals used in research and encouraging signing parties to use animals only in the absence of an alternative, the Convention established and set the basic principles for when and how experiments were to be carried out on animals and also provided technical details on how to house experimental animals. Later in 1986, the Convention was legally shaped by European Economic Community into Directive 86/609/EEC, to improve the control over the use of laboratory animals and to set minimum standards for animals housing and care, as well as for the training of personnel handling these animals and supervising the experiments. It became clear that the 1986 Directive was unsuccessful in creating the EU common framework, mainly due to the discrepancy in the regulation of animal experimentation across the EU. As a result, Directive 2010/63/EU has been introduced in 2010, introducing substantial changes compared to the 86/609/EEC in accordance with the most extensive and demanding national legislations.

Introduction

The history of European legislation on the use of laboratory animals in some countries dates back to the 19th century, for example, Denmark in

1891, Germany in 1883, and Sweden in 1944 [1,2]. However, compared to the rest of Europe, England was a leader in the idea of protecting not only experimental but animals in general. The Royal Society for the Prevention of Cruelty to Animals was founded in London in 1824, and it will soon mark two centuries of existence[1]. The first University of London Animal Welfare Society was founded in 1926 by Charles Hume (1886-1981), who wrote the first textbook in the field and inspired his students, William Russell and Rex Burch to set the principles of 3R (see below).

Today's National Anti-Vivisection Society[2] of England is the first citizens' association to launch a comprehensive campaign against animal experiments[3]. The Society was founded in 1875 in London by Frances Power Cobbe (1822–1904) under the name Victoria Street Society. At that time, about 300 animal experiments were performed in England each year. Through the efforts of this organization, in July of 1875, the English government appointed the First Royal Vivisection Commission which then recommended the adoption of special laws to control vivisection. This led to the adoption of the Cruelty to Animals Act in 1876, as the historically first legal act in this field. This act remained in force for 110 years until it was replaced by the Animals (Scientific Procedures) Act in 1986, as a new national law of England harmonized with the first EU umbrella law in this field, Directive 86/609/EEC.

It is interesting to list a few elements from the Animal Cruelty Act that show the extent to which this legal act was modern at the time it was created (19th century):

- No examination may be performed in accordance with the Law, except to improve existing or develop new knowledge in physiology or knowledge that will be useful for saving or prolonging life or alleviating suffering, or to acquire such knowledge through medical lectures, schools, hospitals, colleges or elsewhere;

---

[1] https://www.rspca.org.uk/

[2] https://www.ufaw.org.uk/

[3] www.navs.org.uk

- The use of experiments on animals for teaching purposes is allowed only if the adequate permit is issued (as described below);

- Each person that performs animal experiments must have the permission of the Minister of Health. In order to obtain a permit, one guarantees that all experiments would be performed at a registered site approved by the Minister and that these sites will be inspected from time to time to ensure compliance with the law;

- Animals must be under the anaesthesia of sufficient strength at all times to prevent pain sensation originating from the experimental procedure. If there is a presumption that the pain would continue after the effect of anaesthetic has ceased or if the animal has been seriously injured, it must be killed before recovering from anaesthesia;

- Licensees are obliged to keep written records of their experiments and to send an annual report to the Minister on the number and nature of experiments performed, as well as to submit additional reports from time to time if necessary.

The basic postulates on which Act was based were actually defined by the British physiologist Marshall Hall (1790–1857): "never do an experiment if the results can be obtained by observation; before starting the experiment, set a clearly defined and achievable goal and gather knowledge about previous similar research; the suffering of experimental animals must be kept to a minimum, and the results of the experiments should be as clear as possible in order to avoid unnecessary repetition". These postulates are woven into all modern laws regarding the welfare of experimental animals.

In Europe several documents and laws have been passed over time to regulate the use of animals in experiments. In 1985 an agreement has been reached in Strasbourg on the Convention on the Protection of Vertebrate Animals Used for Experimental and Other Scientific Purposes (ETS123), being opened for signature by the member states of the Council of Europe on 18. March 1986. So far, it has been signed by the 28 members of the Council of Europe and ratified by 22 members[4]. With the primary aim of

---

[4]   https://www.coe.int/en/web/conventions/full-list?module=signatures-by-treaty&treatynum=123

reducing a number of animals used in research and encouraging signing parties to use animals only in the absence of an alternative, the Convention established and set the basic principles for when and how experiments were to be carried out on animals and also provided technical details on how to house experimental animals [3]. Later in 1986, the Convention was legally shaped by EEC (now the EU) into Directive 86/609/EEC, to improve the control over the use of laboratory animals and to set minimum standards for animals housing and care, as well as for the training of personnel handling these animals and supervising the experiments [4]. In 2003 Directive 86/609/EEC underwent amendment into Directive 2003/65/ EC, with the changes related to the accelerated procedure for revising technical annexes in accordance with new scientific knowledge.

Given their common origin, it is not surprising that the Convention and the Directive overlapped considerably in content, but with the different legal statuses. Conventions are legally banded only to the parties that ratify them, whereas Directives must be implemented by all EU member states. Hence, per definition, the functioning of the former European Communities limited the scope of Directive 86/609/EEC to areas of economic activity (thus excluding animal use within academic research and teaching), whereas Convention ETS123 covered all uses of animals for experimental and other scientific purposes. This discrepancy was partly overcome in 1999 when the EU became a party to Convention ETS123 [3]. However, an important driver in further legislative development was the formation of the EU instead of the EEC. Upon the formation of the European Union in 1993, the EEC was incorporated into the EU and renamed into European Community (EC). In 2009, the EC formally ceased to exist, and its institutions were directly absorbed by the EU. Hence, partly due to the wider political mandate of the EU to harmonize member states regarding the functioning of the single market, but also to prevent member states to develop advantages over others by allowing weaker standards than the agreed minimum when it comes to protecting animal welfare [3], a revision process on the laboratory animals' legislation started. This process first began in 1997 under the Treaty of Amsterdam which required that animal welfare must be embedded in the European legislation. In 1998, through Decision 1999/575/EC Council of Europe, of which the EU is a party, adopted ETS123, thus acknowledging the importance of this field to the EU. Last but not least, the European Centre for the Validation of Alternative Methods (ECVAM) was established in 1991 and helped shape the field.

It became clear that the 1986 Directive was unsuccessful in creating the EU common framework, mainly due to the discrepancy in the regulation of animal experimentation across the EU – while some member states had more extensive national legislation others had merely transposed the 1986 Directive [3]. The revision was requested in 2002 by the European Parliament, challenging the moral and ethical basis for the previous legislation due to the changes in science and public attitudes since Directive 86/609/EEC came into force. The revision process started with the establishment of a Technical Expert Working Group, which brought together national authorities and a wide range of stakeholders to provide scientific and technical advice in response to specific questions raised by the European Commission. At the same time, the Federation of European Laboratory Animal Science Associations (FELASA) independently established a Working Group on Ethical Evaluation of Animal Experimentation aiming to provide unified guidance on how best to conduct the ethical review process within different institutions and countries in Europe, in light of wider societal demand and interest. Accordingly, in its 2005 report, FELASA described and explored a set of principles for conducting the ethical review of laboratory animal use, resulting from a questionnaire that elicited information on how each of the 20 countries represented in FELASA approached such ethical review [5]. In response, the European Commission further sought scientific expert input regarding the animal sentience, the origin of experimental animals, and euthanasia methods, announcing the Community Animal Welfare Action Plan in 2006. Of note, FELASA's role in this area has not been limited to this report only, since it still has an important role regarding different guidelines and recommendations [3,6,7].

The scientific community, industry, and NGOs also followed the revision process closely. Many hearings and policy briefings were organized and statements of several important European research organizations were published till the first public draft version of the revised Directive faced considerable criticism by the research community in fear of research limitations [3]. This debate was coupled with an intense political battle over the new draft version of the Directive in the European Institutions. The turbulent process, described in Nature journal as "more than a decade of pitched battles between research advocates and animal-rights campaigners" [8], finally led to the acceptance of Directive 2010/63/EU. Introducing substantial changes compared to the 86/609/EEC, it settled the standard according to the most extensive and demanding national legislations. In his comparative analysis between the 1986 and the 2010 Directives, Thomas Hartung concluded that little would change for the countries with more demanding legislation, whereas the countries that based the legislation mainly on the Directive 86/609/EEC were to face several new demands [9]. Some of the most important features introduced by Directive 2010/63/EU included: extended scope, enhanced focus on the 3Rs and alternative methods, mandatory project evaluation, severity classification and retrospective assessment, institutional animal welfare bodies, and revised guidelines for accommodation and care, which are now mandatory. To help member states with the implementation process and to ensure a common understanding, the European Commission produced several of guidance documents on different aspects of the Directive, such as the Severity assessment framework, Animal Welfare Bodies and National Committees, Inspections and Enforcement, Genetically altered animals, Project Evaluation and Retrospective Assessment, Availability of Information on the Three Rs, Education and training framework, Non-Technical Project summaries and Specific articles in Directive 2010/63/EU.

Directive 2010/63/EU vs 86/609/EEC

Directive 2010/63/EU is organized into three main sections: 56 recitals, 66 articles, and 7 annexes. The recitals form a structured narrative text displaying reasons for the legal act and guidance for the articles' implementation, whereas the articles and the annexes provide the actual norms or rules introduced by the act [3].

The articles are organized into six chapters. The first chapter, titled General provisions, defines the situations to which the directive applies and lists the purposes for which animals can be used. It also defines key terms and establishes the 3Rs principle as the main guideline in animal experimentation. Chapter II, titled Provision on the use of certain animals for procedures, establishes the limitations in the usage of endangered species, such as nonhuman primates, animals taken from the wild, stray and feral animals of domestic species and defines purpose-bred animals as a standard approach for animals of the typical laboratory species. Chapter III, Procedures, includes more specific provisions in implementing the 3Rs through methods, anaesthesia, severity classification, reuse of animals, endpoints, sharing organs and tissues, and rehoming animals. Chapter IV, Authorization, establishes requirements for breeders, suppliers, and users of animals in terms of conditions of care, responsibilities, and training of personnel and advisory bodies. It further addresses inspections and requirements for project evaluation and authorization. Chapter V, Avoidance of duplication and alternative approaches, focuses on the recognition and development of non-animal alternatives. Chapter VI, Final provisions, sets the rules for implementation, adaptation, reporting, and the role of different entities. Finally, the seven annexes provide more detailed information complementing some of the articles [3].

The first seven recitals of 56 in total, reason the need for the new Directive. Recital 1 stresses the weakness of the previous Directive in the light of common EU framework development, stating that "On 24 November 1986 the Council of the European Union adopted Directive 86/609/ EEC (3) in order to eliminate disparities between laws, regulations and administrative provisions of the Member States regarding the protection of animals used for experimental and other scientific purposes. Since the adoption of that Directive, further disparities between Member States have emerged. Certain Member States have adopted national implementing measures that ensure a high level of protection of animals used for scientific purposes, while others only apply the minimum requirements laid down in Directive 86/609/EEC. These disparities are liable to constitute barriers to trade in products and substances the development of which involves experiments on animals. Accordingly, this Directive should provide for more detailed rules in order to reduce such disparities by approximating the rules applicable in that area and to ensure a proper functioning of the internal market". Recitals 2 to 4 give a historical viewpoint on the decisions preceding the revision process of the Directive 86/609/EEC. Recital 2 underlines the value of animal welfare to the Union, as enshrined in Article 13 of the Treaty on the Functioning of the European Union (TFEU), stating that "in formulating and implementing the Union's agriculture, fisheries, transport, internal market, research and technological development and space policies, the Union and the Member States shall, since animals are sentient beings, pay full regard to the welfare requirements of animals, while respecting the legislative or administrative provisions and customs of the Member States relating in particular to religious rites, cultural traditions and regional heritage". Recital 3 emphasizes the importance of adopting Decision 1999/575/EC by the Council of the European Union and Community, whereas recital 4 reminds of the European Parliament resolution that called for the Commission to come forward with a proposal for a revision of Directive 86/609/EEC with more stringent and transparent measures in the area of animal experimentation. Recital 6 points out the necessity "to improve the welfare of animals used in scientific procedures by raising the minimum standards for their protection in line with the latest scientific developments", while recital 7 deals with the flexibility of member state national legislation toward standards defined by the Directive. Being a legislative umbrella for the EU, the Directive sets the minimum standards that need to be achieved by all member states, but also allows them to embed in national legislative "more extensive animal-welfare rules than those agreed upon at the level of the Union, so far as they are compatible with the TFEU".

The remaining recitals deal with the most important Directive 2010/63/EU elements, some of which represent novelties in regard to the Directive 86/609/EEC (Table 1).

Table 1. Some differences and novelties by comparing Directives 86/609/EEC and 2010/63/EU.

Directive 86/609/EEC
Article 1. The aim of this Directive is to ensure that where animals are used for experimental or other scientific purposes the provisions laid down by law, regulation or administrative provisions in the Member States for their protection are approximated so as to avoid affecting the establishment and functioning of the common market, in particular by distortions of competition or barriers to trade.
Directive 2010/63/EU
Article 1. 1.   This Directive establishes measures for the protection of animals used for scientific or educational purposes. To that end, it lays down rules on the following: (a) the replacement and reduction of the use of animals in procedures and the refinement of the breeding, accommodation, care and use of animals in procedures; (b) the origin, breeding, marking, care and accommodation and killing of animals; (c) the operations of breeders, suppliers and users; (d) the evaluation and authorisation of projects involving the use of animals in procedures. 2.   This Directive shall apply where animals are used or intended to be used in procedures, or bred specifically so that their organs or tissues may be used for scientific purposes. This Directive shall apply until the animals referred to in the first subparagraph have been killed, rehomed or returned to a suitable habitat or husbandry system.
Comments:
1.   The scope of the animals' protection has shifted from the issue of the competition and barriers to trade in the common market to the application of the 3R principle, reflecting both EEC and EC limitations to establish legislation outside the economy framework; 2.   The scope of the animals' protection now includes not only "experimental or other scientific purposes", but also educational; 3.   By specifically defining breeders, suppliers and users, the new Directive encompasses all aspects of animal use, not only the scientific part. The consequence is that an appropriate licence must be obtained before the start of the experiment. Therefore, for all animals housed in the institution, a previously obtained permit must be displayed in case of inspection; 4.   The evaluation and authorisation of projects involving the use of animals in procedures have been introduced.
Directive 86/609/EEC
Article 2. For the purposes of this Directive the following definitions shall apply: (a) 'animal' unless otherwise qualified, means any live non-human vertebrate, including free-living larval and/or reproducing larval forms, but excluding foetal or embryonic forms.

Directive 2010/63/EU

Article 1.

3.   This Directive shall apply to the following animals :

  (a) live non-human vertebrate animals, including:

  i) independently feeding larval forms, and ii) foetal forms of mammals as from the last third of their normal development;

4.   (b) live cephalopods.

Comments:

1.   A list of experimental animals was extended to include foetal organisms and cephalopods.

Directive 86/609/EEC

Article 2.

For the purposes of this Directive the following definitions shall apply:

(b) 'experimental animals ' means animals used or to be used in experiments;

(d) 'experiment' means any use of an animal for experimental or other scientific purposes which may cause it pain, suffering, distress or lasting harm, including any course of action intended, or liable, to result in the birth of an animal in any such condition, but excluding the least painful methods accepted in modern practice (i.e. 'humane' methods) of killing or marking an animal ; an and ends when no further observations are to be made for that experiment; the elimination of pain, suffering, distress or lasting harm by the successful use of anaesthesia or analgesia or other methods does not place the use of an animal outside the scope of this definition. Non-experimental, agricultural, or clinical veterinary practices are excluded.

Directive 2010/63/EU

Article 3.

For the purposes of this Directive the following definitions shall apply:

1.   'procedure' means any use, invasive or non-invasive, of an animal for experimental or other scientific purposes, with known or unknown outcome, or educational purposes, which may cause the animal a level of pain, suffering, distress or lasting harm equivalent to, or higher than, that caused by the introduction of a needle in accordance with good veterinary practice. This includes any course of action intended, or liable, to result in the birth or hatching of an animal or the creation and maintenance of a genetically modified animal line in any such condition, but excludes the killing of animals solely for the use of their organs or tissues;

2.   'project' means a programme of work having a defined scientific objective and involving one or more procedures.

Article 1.

5.   This Directive shall not apply to the following:

  (a) non-experimental agricultural practices;

  (b) non-experimental clinical veterinary practices;

  (c) veterinary clinical trials required for the marketing authorisation of a veterinary medicinal product;

(d) practices undertaken for the purposes of recognised animal husbandry;

(e) practices undertaken for the primary purpose of identification of an animal;

(f) practices not likely to cause pain, suffering, distress or lasting harm equivalent to, or higher than, that caused by the introduction of a needle in accordance with good veterinary practice.

Comments:

1. A term "experiment" has been expanded and clarified as a "procedure" and "project". An important consequence is the simplification of the evaluation and authorisation process making it possible to obtain a single license for complex projects that include a number of different experiments, species, and could last for years;

2. The list of procedures that are excluded from the Directive is also expanded and clarified, and now includes not only non-experimental practices, but also scientific "practices not likely to cause pain, suffering, distress or lasting harm equivalent to, or higher than, that caused by the introduction of a needle in accordance with good veterinary practice".

Directive 86/609/EEC

Article 3.

This Directive applies to the use of animals in experiments which are undertaken for one of the following purposes:

(a) the development, manufacture, quality, effectiveness and safety testing of drugs, foodstuffs and other substances or products:

(i) for the avoidance, prevention, diagnosis or treatment of disease, ill-health or other abnormality or their effects in man, animals or plants;

(ii) for the assessment, detection, regulation or modification of physiological conditions in man, animals or plants;

(b) the protection of the natural environment in the interests of the health or welfare of man or animal.

Directive 2010/63/EU

Article 5.

Procedures may be carried out for the following purposes only:

(a) basic research;

(b) translational or applied research with any of the following aims:

(i) the avoidance, prevention, diagnosis or treatment of disease, ill-health or other abnormality or their effects in human beings, animals or plants;

(ii) the assessment, detection, regulation or modification of physiological conditions in human beings, animals or plants; or

(iii) the welfare of animals and the improvement of the production conditions for animals reared for agricultural purposes.

(c) for any of the aims in point (b) in the development, manufacture or testing of the quality, effectiveness and safety of drugs, foodstuffs and feed-stuffs and other substances or products;

(d)	protection of the natural environment in the interests of the health or welfare of human beings or animals;
(e)	research aimed at preservation of the species;
(f)	higher education, or training for the acquisition, maintenance or improvement of vocational skills;
(g)	forensic inquiries.

Comments:

1.	The list of purposes in which animals are used has been extended for the preservation of the species, education/vocational skills, and forensic inquiries. However, as stated in Articles 7 and 8, this list is further limited in the case of endangered species and non-human primates.

Directive 86/609/EEC

None.

Directive 2010/63/EU

2.	In choosing between procedures, those which to the greatest extent meet the following requirements shall be selected:
	(b) involve animals with the lowest capacity to experience pain, suffering, distress or lasting harm.

Comments:

1.	The term "capacity to experience pain, suffering, distress or lasting harm" was introduced, which led to (more or less justified) division into animals with lower and higher capacities.

Directive 86/609/EEC

None.

Directive 2010/63/EU

Article 13.

3.	Death as the end-point of a procedure shall be avoided as far as possible and replaced by early and humane end-points. Where death as the end-point is unavoidable, the procedure shall be designed so as to:
	(a) result in the deaths of as few animals as possible; and
	(b) reduce the duration and intensity of suffering to the animal to the minimum possible and, as far as possible, ensure a painless death.

Comments:

1.	Avoidance of death as an endpoint (as far as possible) has been introduced.

Directive 86/609/EEC

None.

Directive 2010/63/EU

Article 15.

Classification of severity of procedures

1. Member States shall ensure that all procedures are classified as 'non-recovery', 'mild', 'moderate', or 'severe' on a case-by-case basis using the assignment criteria set out in Annex VIII.

2. Subject to the use of the safeguard clause in Article 55(3), Member States shall ensure that a procedure is not performed if it involves severe pain, suffering or distress that is likely to be long-lasting and cannot be ameliorated.

Comments:

1. An entirely new concept of severity has been introduced, with the ban on very severe and long-lasting procedures that cannot be ameliorated.

Directive 86/609/EEC

None.

Directive 2010/63/EU

Article 18.

Member States shall facilitate, where appropriate, the establishment of programmes for the sharing of organs and tissues of animals killed.

Comments:

1. Sharing of organs and tissues has been introduced as a way of animal use reduction.

Directive 86/609/EEC

Article 3.

(f) 'competent person' means any person who is considered by a Member State to be competent to perform the relevant function described in this Directive.

Directive 2010/63/EU

Article 23.

Competence of personnel

1. Member States shall ensure that each breeder, supplier and user has sufficient staff on site.

2. The staff shall be adequately educated and trained before they perform any of the following functions:

   (a) carrying out procedures on animals;

   (b) designing procedures and projects;

   (c) taking care of animals; or

   (d) killing animals.

Persons carrying out the functions referred to in point (b) shall have received instruction in a scientific discipline relevant to the work being undertaken and shall have species-specific knowledge.

Staff carrying out functions referred to in points (a), (c) or (d) shall be supervised in the performance of their tasks until they have demonstrated the requisite competence.

Member States shall ensure, through authorisation or by other means, that the requirements laid down in this paragraph are fulfilled.

3.   Member States shall publish, on the basis of the elements set out in Annex V, minimum requirements with regard to education and training and the requirements for obtaining, maintaining and demonstrating requisite competence for the functions set out in paragraph 2.

4.   Non-binding guidelines at the level of the Union on the requirements laid down in paragraph 2 may be adopted in accordance with the advisory procedure referred to in Article 56(2).

Article 25.

Member States shall ensure that each breeder, supplier and user has a designated veterinarian with expertise in laboratory animal medicine, or a suitably qualified expert where more appropriate, charged with advisory duties in relation to the well-being and treatment of the animals.

Comments:

1.   The concept of "competent persons" and their competencies and roles has been strongly expanded and clarified.

Directive 86/609/EEC

None.

Directive 2010/63/EU

Article 34.

Inspections by the Member States

1.   Member States shall ensure that the competent authorities carry out regular inspections of all breeders, suppliers and users, including their establishments, to verify compliance with the requirements of this Directive.

2.   The competent authority shall adapt the frequency of inspections on the basis of a risk analysis for each establishment, taking account of:

(a) the number and species of animals housed;

(b) the record of the breeder, supplier or user in complying with the requirements of this Directive;

(c) the number and types of projects carried out by the user in question; and

(d) any information that might indicate non-compliance.

3.   Inspections shall be carried out on at least one third of the users each year in accordance with the risk analysis referred to in paragraph 2. However, breeders, suppliers and users of non- human primates shall be inspected at least once a year.

4.   An appropriate proportion of the inspections shall be carried out without prior warning.

5.   Records of all inspections shall be kept for at least 5 years.

Comments:

1. The scope of inspections and their roles have been defined.

Directive 86/609/EEC

Article 12.

1. Member States shall establish procedures whereby experiments themselves or the details of persons conducting such experiments shall be notified in advance to the authority.

2. Where it is planned to subject an animal to an experiment in which it will, or may, experience severe pain which is likely to be prolonged, that experiment must be specifically declared and justified to, or specifically authorized by the authority. The authority shall take appropriate judicial or administrative action if it is not satisfied that the experiment is of sufficient importance for meeting the essential needs of man or animal.

Article 24.

This Directive shall not restrict the right of the Member States to apply or adopt stricter measures for the protection of animals used in experiments or for the control and restriction of the use of animals for experiments. In particular, Member States may require a prior authorization for experiments or programmes of work notified in accordance with the provisions of Article 12 (1).

Directive 2010/63/EU

Article 36.

1. Member States shall ensure, without prejudice to Article 42, that projects are not carried out without prior authorisation from the competent authority, and that projects are carried out in accordance with the authorisation or, in the cases referred to in Article 42, in accordance with the application sent to the competent authority or any decision taken by the competent authority.

2. Member States shall ensure that no project is carried out unless a favourable project evaluation by the competent authority has been received in accordance with Article 38.

Article 37.

1. Member States shall ensure that an application for project authorisation is submitted by the user or the person responsible for the project. The application shall include at least the following:

   (a) the project proposal;

   (b) a non-technical project summary; and

   (c) information on the elements set out in Annex VI.

2. Member States may waive the requirement in paragraph 1(b) for projects referred to in Article 42(1).

Article 38.

1. The project evaluation shall be performed with a degree of detail appropriate for the type of project and shall verify that the project meets the following criteria:

   (a) the project is justified from a scientific or educational point of view or required by law;

(b) the purposes of the project justify the use of animals; and

(c) the project is designed so as to enable procedures to be carried out in the most humane and environmentally sensitive manner possible.

2.   The project evaluation shall consist in particular of the following:

(a) an evaluation of the objectives of the project, the predicted scientific benefits or educational value;

(b) an assessment of the compliance of the project with the requirement of replacement, reduction and refinement;

(c) an assessment and assignment of the classification of the severity of procedures;

(d) a harm-benefit analysis of the project, to assess whether the harm to the animals in terms of suffering, pain and distress is justified by the expected outcome taking into account ethical considerations, and may ultimately benefit human beings, animals or the environment;

(e) an assessment of any justification referred to in Articles 6 to 12, 14, 16 and 33; and

(f) a determination as to whether and when the project should be assessed retrospectively.

3.   The competent authority carrying out the project evaluation shall consider expertise in particular in the following areas:

(a) the areas of scientific use for which animals will be used including replacement, reduction and refinement in the respective areas;

(b) experimental design, including statistics where appropriate;

(c) veterinary practice in laboratory animal science or wildlife veterinary practice where appropriate;

(d) animal husbandry and care, in relation to the species that are intended to be used.

4.   The project evaluation process shall be transparent.

Subject to safeguarding intellectual property and confidential information, the project evaluation shall be performed in an impartial manner and may integrate the opinion of independent parties.

Article 39.

1.   Member States shall ensure that when determined in accordance with Article 38(2)(f), the retrospective assessment shall be carried out by the competent authority which shall, on the basis of the necessary documentation submitted by the user, evaluate the following:

(a) whether the objectives of the project were achieved;

(b) the harm inflicted on animals, including the numbers and species of animals used, and the severity of the procedures; and

(c) any elements that may contribute to the further implementation of the requirement of replacement, reduction and refinement.

2.   All projects using non-human primates and projects involving procedures classified as 'severe', including those referred to in Article 15(2), shall undergo a retrospective assessment.

3.   Without prejudice to paragraph 2 and by way of derogation from Article 38(2)(f), Member States may exempt projects involving only procedures classified as 'mild' or 'non- recovery' from the requirement for a retrospective assessment.

Article 40.

1. The project authorisation shall be limited to procedures which have been subject to:

    (a) a project evaluation; and

    (b) the severity classifications assigned to those procedures.

2. The project authorisation shall specify the following:

    (a) the user who undertakes the project;

    (b) the persons responsible for the overall implementation of the project and its compliance with the project authorisation;

    (c) the establishments in which the project will be undertaken, where applicable; and

    (d) any specific conditions following the project evaluation, including whether and when the project shall be assessed retrospectively.

3. Project authorisations shall be granted for a period not exceeding 5 years.

4. Member States may allow the authorisation of multiple generic projects carried out by the same user if such projects are to satisfy regulatory requirements or if such projects use animals for production or diagnostic purposes with established methods.

Article 41.

1. Member States shall ensure that the decision regarding authorisation is taken and communicated to the applicant 40 working days at the latest from the receipt of the complete and correct application. This period shall include the project evaluation.

2. When justified by the complexity or the multi-disciplinary nature of the project, the competent authority may extend the period referred to in paragraph 1 once, by an additional period not exceeding 15 working days. The extension and its duration shall be duly motivated and shall be notified to the applicant before the expiry of the period referred to in paragraph 1.

3. Competent authorities shall acknowledge to the applicant all applications for authorisations as quickly as possible, and shall indicate the period referred to in paragraph 1 within which the decision is to be taken.

4. In the case of an incomplete or incorrect application, the competent authority shall, as quickly as possible, inform the applicant of the need to supply any additional documentation and of any possible effects on the running of the applicable time period.

Article 42.

1. Member States may decide to introduce a simplified administrative procedure for projects containing procedures classified as 'non-recovery', 'mild' or 'moderate' and not using non-human primates, that are necessary to satisfy regulatory requirements, or which use animals for production or diagnostic purposes with established methods.

Article 43.

1. Subject to safeguarding intellectual property and confidential information, the non-technical project summary shall provide the following:

    (a) information on the objectives of the project, including the predicted harm and benefits and the number and types of animals to be used;

    (b) a demonstration of compliance with the requirement of replacement, reduction and refinement.

The non-technical project summary shall be anonymous and shall not contain the names and addresses of the user and its personnel.

1. Member States may require the non-technical project summary to specify whether a project is to undergo a retrospective assessment and by what deadline. In such a case, Member States shall ensure that the non-technical project summary is updated with the results of any retrospective assessment.

2. Member States shall publish the non-technical project summaries of authorised projects and any updates thereto.

Article 44.

1. Member States shall ensure that amendment or renewal of the project authorisation is required for any change of the project that may have a negative impact on animal welfare.

2. Any amendment or renewal of a project authorisation shall be subject to a further favourable outcome of the project evaluation.

3. The competent authority may withdraw the project authorisation where the project is not carried out in accordance with the project authorisation.

4. Where a project authorisation is withdrawn, the welfare of the animals used or intended to be used in the project must not be adversely affected.

5. Member States shall establish and publish conditions for amendment and renewal of project authorisations.

Article 45.

1. Member States shall ensure that all relevant documentation, including project authorisations and the result of the project evaluation is kept for at least 3 years from the expiry date of the authorisation of the project or from the expiry of the period referred to in Article 41(1) and shall be available to the competent authority.

2. Without prejudice to paragraph 1, the documentation for projects which have to undergo retrospective assessment shall be kept until the retrospective assessment has been completed.

Comments:

1. The scope of project authorisation has been strongly expanded and clarified. It now includes a clear timeframe of all authorisation phases, including application, evaluation, decision, amendment, renewal, withdrawal, and documentation;

2. A simplified administrative procedure has been introduced regarding project severity classification;

3. Publication of anonymous non-technical project summaries has been introduced;

4. Retrospective assessment of projects involving procedures with severe harm, projects involving non-human primates, and projects selected within the evaluation of applications have been introduced.

Directive 86/609/EEC

Article 23

1. The Commission and Member States should encourage research into the development and validation of alternative techniques which could provide the same level of information as that obtained in experiments using animals, but which involve fewer animals, or which entail less painful procedures, and shall take such other steps as they consider appropriate to encourage research in this field. The Commission and Member States shall monitor trends in experimental methods.

Directive 2010/63/EU

Article 47

1. The Commission and the Member States shall contribute to the development and validation of alternative approaches which could provide the same or higher levels of information as those obtained in procedures using animals, but which do not involve the use of animals or use fewer animals, or which entail less painful procedures, and they shall take such other steps as they consider appropriate to encourage research in this field.

2. Member States shall assist the Commission in identifying and nominating suitable specialised and qualified laboratories to carry out such validation studies.

3. After consulting the Member States, the Commission shall set the priorities for those validation studies and allocate the tasks between the laboratories for carrying out those studies.

4. Member States shall, at national level, ensure the promotion of alternative approaches and the dissemination of information thereon.

5. Member States shall nominate a single point of contact to provide advice on the regulatory relevance and suitability of alternative approaches proposed for validation.

6. The Commission shall take appropriate action with a view to obtaining international acceptance of alternative approaches validated in the Union.

Article 48

1. The Union Reference Laboratory and its duties and tasks shall be those referred to in Annex VII.

2. The Union Reference Laboratory may collect charges for the services it provides that do not directly contribute to the further advancement of replacement, reduction and refinement.

3. Detailed rules necessary for the implementation of paragraph 2 of this Article and Annex VII may be adopted in accordance with the regulatory procedure referred to in Article 56(3).

Comments:

1. The importance of alternative approaches is underlined, with the clarification of their development, validation, dissemination, and international acceptance;

2. The Union Reference Laboratory has been introduced through the following tasks (Annex VII):

   (a) coordinating and promoting the development and use of alternatives to procedures including in the areas of basic and applied research and regulatory testing;

   (b) coordinating the validation of alternative approaches at Union level;

   (c) acting as a focal point for the exchange of information on the development of alternative approaches;

   (d) setting up, maintaining and managing public databases and information systems on alternative approaches and their state of development;

   (e) promoting dialogue between legislators, regulators, and all relevant stakeholders, in particular, industry, biomedical scientists, consumer organisations and animal-welfare groups, with a view to the development, validation, regulatory acceptance, international recognition, and application of alternative approaches.

Directive 86/609/EEC
None.

Directive 2010/63/EU
Article 49
1.   Each Member State shall establish a national committee for the protection of animals used for scientific purposes. It shall advise the competent authorities and animal-welfare bodies on matters dealing with the acquisition, breeding, accommodation, care and use of animals in procedures and ensure sharing of best practice.
2.   The national committees referred to in paragraph 1 shall exchange information on the operation of animal-welfare bodies and project evaluation and share best practice within the Union.

Comments:
1.   The concept of national committees has been introduced.

Directive 2010/63/EU and beyond – Expanding the 3R principles

Looking ahead, there are great opportunities for the advancement of the 3R principles [10]. The concept of replacement holds a special place among the 3Rs. It was first introduced by Russell and Burch, thus reflecting the intended order in which the 3Rs were to be considered: a principle of reduction has sense only if the replacement was first considered and excluded. Replacement is also well accepted by both researchers and the wider community as it is the only concept of the 3Rs fully compatible with the animal rights perspective. Hence, new technological options, such as computer modelling, human cell and tissue cultures, and sophisticated imaging and analysis will strongly contribute to the full implementation of the replacement.

Data and/or tissues sharing, highlighted in Directive 2010/63/EU, may contribute to the animal use reduction [10]. In addition, the number of animals used can be reduced through the improvement of experimental design and statistical approaches [11]. An analysis of the 53 'landmark' papers in the preclinical cancer field found that 89% percent of the studies failed to be reproduced due to the poor study design, investigator bias, and incomplete reporting [12]. This calls for more rigorous methodological approaches, such as randomization and blinding, to reduce bias and improve research validity. Another specific approach to the reduction of animal use includes avoidance of candidate drug attrition [13]. Namely, a high percentage of candidate drugs never reach a clinical trial due to a lack of efficacy or safety, previously not predicted in animal tests [14-16]. Various initiatives, such as Europe's Innovative Medicines Initiative and the FDA's Critical Path Initiative, are underway to develop new methods for the early screening of drug failures and the selection of drug candidates that are most likely to succeed. Similarly, there is concern regarding the use of animals in environmental pollutants testing. The results interpretation and extrapolation to the human population are often difficult since experimental animals are usually exposed to doses much higher than those typical for human exposures [17]. That is why organizations such as the National Research Council and Organisation for Economic Co-operation and Development called for the development of more predictive mechanism-based assays, with an increased focus on *in vitro* and *in silico* approaches.

A concept of refinement calls for the implementation of training protocols that help improve animal handling thus reducing the stress that arises from the experimental procedures [10]. Revision and improvement of scoring systems are needed to better assess animal welfare [18]. Environmental enrichment of animal housing is now obligatory as specified by Directive 2010/63/EU (see Annex III: Requirements for establishments and for the care and accommodation of animals), as well as by codes of practice and regulations in many countries [19], allowing animals the range of natural activities, including physical exercise, foraging, and manipulative and cognitive activities [10].

Regardless of the 3Rs, there are additional questions that are to be tackled in the future. For instance, the idea that there is an ethical gain in moving from "higher" to "lower" organisms was originally referred to by Russell and Burch as "comparative replacement". Today it is explicit in Directive 2010/63/EU, requiring scientists to choose procedures that "involve animals with the lowest capacity to experience pain, suffering, distress or lasting harm" (Article 13.2). Nevertheless, even if the concept does seem intuitively correct, there is a problem with its accuracy. There is still no clear-cut way of defining the capacity of an organism to experience pain, suffering, or distress. Attempts have been made to define criteria for larger groups of species [20]. Current rankings relate more to the socio-zoological scale [21], i.e., how animals are perceived by humans, than to the real animals' ability to suffer. Therefore, the interpretation and implementation of the 3Rs principles are partly dependent on the investigator's moral principles.

Finally, by improving scientific reliability, validity and reproducibility, global standards for reporting and transparency could also be improved [22]. In 2010, the Animal Research Reporting of in vivo experiments (ARRIVE) guidelines has been published by UK National Centre for the Replacement Refinement & Reduction of Animals [23], and further revised in 2020 [24]. Despite numerous journals referring to the ARRIVE guidelines in their "guide to authors", many authors do not adhere to them [22]. A study from 2016 showed that no matter if the journal asked authors to adhere to the ARRIVE publishing guidelines or not, the quality of the methods section was not essentially improved [25]. Hence, one of the solutions could include advising research funders and publishers to encourage or require the investigators to adhere to the ARRIVE guidelines [22]. This approach is already implemented by the US National Institutes of Health[5] and UK Medical Research Council, and journals such as the Nature and PLoS families [23].

References

1. Berry, A.; Vitale, A.; Carere, C.; Alleva, E. EU guidelines for the care and welfare of an "exceptional invertebrate class" in scientific research. Commentary. *Ann Ist Super Sanita* **2015**, 51, 267-269, doi:10.4415/ANN_15_04_04.
2. Hubrecht, R.C. The Welfare of Animals Used in Research : Practice and Ethics. **2014**.
3. Olsson, I.A.S.; Silva, S.P.D.; Townend, D.; Sandoe, P. Protecting Animals and Enabling Research in the European Union: An Overview of Development and Implementation of Directive 2010/63/EU. ILAR J **2016**, 57, 347-357, doi:10.1093/ilar/ilw029.
4. Louhimies, S. Directive 86/609/EEC on the protection of animals used for experimental and other scientific purposes. *Altern Lab Anim* **2002**, *30 Suppl 2*, 217-219, doi:10.1177/026119290203002S36.
5. Smith, J.A.; van den Broek, F.A.; Martorell, J.C.; Hackbarth, H.; Ruksenas, O.; Zeller, W.; Experiments, F.W.G.o.E.E.o.A. Principles and practice in ethical review of animal experiments across Europe: summary of the report of a FELASA working group on ethical evaluation of animal experiments. *Lab. Anim.* **2007**, *41*, 143-160, doi:10.1258/002367707780378212.

---

5 https://grants.nih.gov/reproducibility/index.htm

6. Guillen, J. FELASA guidelines and recommendations. *J Am Assoc Lab Anim Sci* **2012**, *51*, 311-321.

7. Bronstad, A.; Newcomer, C.E.; Decelle, T.; Everitt, J.I.; Guillen, J.; Laber, K. Current concepts of Harm-Benefit Analysis of Animal Experiments - Report from the AALAS-FELASA Working Group on Harm-Benefit Analysis - Part 1. *Lab. Anim.* **2016**, *50*, 1-20, doi:10.1177/0023677216642398.

8. Abbott, A. Lab-animal battle reaches truce. *Nature* **2010**, *464*, 964, doi:10.1038/464964a.

9. Hartung, T. Comparative analysis of the revised Directive 2010/6106/EU for the protection of laboratory animals with its predecessor 86/609/EEEEC – a t4 report. *ALTEX - Alternatives to animal experimentation* **2010**, *27*, 285-303, doi:10.14573/altex.2010.4.285.

10. Hubrecht, R.C.; Carter, E. The 3Rs and Humane Experimental Technique: Implementing Change. *Animals (Basel)* **2019**, *9*, doi:10.3390/ani9100754.

11. Howard, B.; Nevalainen, T.; Perretta, G. *The COST manual of laboratory animal care and use: refinement, reduction, and research*; Taylor & Francis: 2010.

12. Begley, C.G.; Ellis, L.M. Drug development: Raise standards for preclinical cancer research. *Nature* **2012**, *483*, 531-533, doi:10.1038/483531a.

13. Graham, M.L.; Prescott, M.J. The multifactorial role of the 3Rs in shifting the harm-benefit analysis in animal models of disease. *Eur J Pharmacol* **2015**, *759*, 19-29, doi:10.1016/j.ejphar.2015.03.040.

14. Kola, I.; Landis, J. Can the pharmaceutical industry reduce attrition rates? *Nat. Rev. Drug Discov.* **2004**, *3*, 711-715, doi:10.1038/nrd1470.

15. Walker, I.; Newell, H. Do molecularly targeted agents in oncology have reduced attrition rates? *Nat. Rev. Drug Discov.* **2009**, *8*, 15-16, doi:10.1038/nrd2758.

16. Bailey, J.; Thew, M.; Balls, M. An analysis of the use of animal models in predicting human toxicology and drug safety. *Altern Lab Anim* **2014**, *42*, 181-199, doi:10.1177/026119291404200306.

17. Leist, M.; Hartung, T.; Nicotera, P. The dawning of a new age of toxicology. *ALTEX* **2008**, *25*, 103-114.

18. Honess, P.; Wolfensohn, S. The extended welfare assessment grid: a matrix for the assessment of welfare and cumulative suffering in experimental animals. *Altern Lab Anim* **2010**, *38*, 205-212, doi:10.1177/026119291003800304.

19. Andre, V.; Gau, C.; Scheideler, A.; Aguilar-Pimentel, J.A.; Amarie, O.V.; Becker, L.; Garrett, L.; Hans, W.; Holter, S.M.; Janik, D., et al. Laboratory mouse housing conditions can be improved using common environmental enrichment without compromising data. *PLoS Biol* **2018**, *16*, e2005019, doi:10.1371/journal.pbio.2005019.

20. Smith, J.A.B.K.M. *Lives in the Balance : the Ethics of Using Animals in Biomedical Research : The Report of a Working Party of the Institute of Medical Ethics*; Oxford University Press: Oxford, 1991.

21. Arluke, A. *Regarding animals*; Temple University Press: [S.l.], 2022.

22. Aske, K.C.; Waugh, C.A. Expanding the 3R principles: More rigour and transparency in research using animals. *EMBO Rep* **2017**, *18*, 1490-1492, doi:10.15252/embr.201744428.

23. Kilkenny, C.; Browne, W.J.; Cuthill, I.C.; Emerson, M.; Altman, D.G. Improving bioscience research reporting: the ARRIVE guidelines for reporting animal research. *PLoS Biol* **2010**, *8*, e1000412, doi:10.1371/journal.pbio.1000412.

24. Percie du Sert, N.; Hurst, V.; Ahluwalia, A.; Alam, S.; Avey, M.T.; Baker, M.; Browne, W.J.; Clark, A.; Cuthill, I.C.; Dirnagl, U., et al. The ARRIVE guidelines 2.0: Updated guidelines for reporting animal research. *PLoS Biol* **2020**, *18*, e3000410, doi:10.1371/journal.pbio.3000410.

25. Carbone, L.; Austin, J. Pain and Laboratory Animals: Publication Practices for Better Data Reproducibility and Better Animal Welfare. *PLoS One* **2016**, *11*, e0155001, doi:10.1371/journal.pone.0155001.

ETHICAL EVALUATION OF ANIMAL RESEARCH PROPOSALS

Alexander M. Trbovich, M.D., Ph.D.

Research Professor of Medicine
Professor of Pathophysiology, University of Belgrade School of Medicine, Belgrade, Serbia

Abstract

The ethics review aims to ensure that the grant recipients, researchers and the host institutions follow the ethical principles and relevant legislation. If this is not the case, the ethics evaluation panel will reject proposals and terminate projects. The approved research should follow the highest ethical standards and be acceptable to the general public. By ensuring these goals, researchers are protected from litigation, controversy, and hassle.

The applicant for the proposal must undertake and pass an accredited training program and be familiar with the legislation, regulation, and ethical framework involved in using animals for scientific purposes. As a part of the application to work with animals, a researcher writes a non-technical summary (NTS). NTS is visible to the general public and helps laypeople understand why animals are used in scientific research and how the researcher will implement the 3Rs policy.

The status of animals varies significantly in different societies. Consequently, the ethics review process varies among local and national ethics committees. In the one-step evaluation, both scientific and ethical parts are evaluated simultaneously. The ethical part is evaluated independently from the scientific part in the two-step evaluation. Each approved proposal must have at least three licenses: a) a personal license for each person carrying out procedures on animals, b) a project license for the approved work, and c) an establishment license for the place at which the work is carried out. The specific purpose of the ethics review of the proposed research is to protect the welfare of animals and the researchers' safety.

The purpose of review

The ethics review aims to ensure that recipients of the EC grants, both the researchers and the host institutions, follow the EU ethics principles. These principles are aligned with the EU and national ethics legislations. The legal basis for the ethics review process is given in the EU Regulation 2021/695, Document 32021R0695 [1]. The ethics evaluation secures that the proposed research follows commonly accepted ethical standards and relevant legislation. If this is not the case, the ethics evaluation panel will reject proposals and terminate projects. The goal of an ethics review is to ensure that the research is conducted to the highest ethical standards and acceptable to the general public, and that the researchers are protected from litigation, controversy, and hassle.

Specific research activities, such as human cloning for reproductive purposes, are excluded from funding. However, some research, for example, on embryonic stem cells, may be financed depending on the contents of the scientific proposal and the legal framework of the EU Member States involved. The EC will not fund research activity banned in that EU Member State. Also, the EC will grant no funding for prohibited research activities in all the EU Member States. If a place of conducting research is non-EU country, the applicant needs to confirm that the same activities would have been allowed in an EU Member State. Regarding the use of experimental animals, the idea is to reduce their number with the ultimate goal of replacing the use of animals in research [1]. The EC grant approving agencies also have inspection systems to ensure that set ethics rules are not violated.

The applicant

The applicant for the proposal must undertake and pass an accredited training program and be familiar with the legislation, regulation, and ethical framework involved in using animals for scientific purposes. When submitting the proposal, the applicant is requested to fill in the Ethics Issues Table and invited to write an ethics self-assessment in Part A of the application form and, if needed, in an additional ethics annex [2]. The applicant must describe foreseeable ethics issues and demonstrate their conformity with national and European regulations. Also, the applicant has to confirm that planned activities will comply with the European Code of Conduct for Research Integrity [3]. In the case of collaboration with or conducting research in resource-poor countries, the applicant must apply the Global Code of Conduct for Research in Resource-Poor Settings [4].

The applicant must obtain all approvals or other mandatory documents from the relevant national, local ethics committees, or other bodies before starting the appropriate activities. Three essential licenses are:

1. A personal license for each person carrying out procedures on animals.
2. A project license for the approved work.
3. An establishment license for the place at which the work is carried out.

The applicant must keep these and all other needed documents on file and, upon request, provide them to the European Commission or the relevant funding body.

The ARRIVE and PREPARE guidelines

The approval for the proposed animal experiments from the local ethics committee significantly facilitates approvals from the national or other higher authorities if they are needed. The ARRIVE and PREPARE guidelines help in a successful application for ethics approval.

The ARRIVE guidelines 2.0 [5,6,7]

The ARRIVE (Animal Research: Reporting of In Vivo Experiments) guidelines 2.0 are a checklist of information for publications describing animal research to ensure that studies are reported in enough detail to add to the knowledge base. It consists of the ARRIVE Essential 10, the Recommended Set, and Glossary. The ARRIVE Essential 10 and the Recommended Set consist of explanations and provided examples.

The ARRIVE Essential 10 consists of the following subcategories: study design, sample size, inclusion and exclusion criteria, randomization, blinding, outcomes measures, statistical methods, experimental animals, experimental procedures, and results.

The Recommended set includes additional 11 subcategories: abstract, background, objectives, ethical statement, housing and husbandry, animal care and monitoring, interpretation/scientific implications, generalizability/translation, protocol registration, data access, and declaration of interests.

The PREPARE guidelines [8]

The PREPARE (Planning Research and Experimental Procedures on Animals: Recommendations for Excellence) guidelines for planning experiments are developed to reduce waste, promote animal alternatives, and increase the reproducibility of research and testing. The PREPARE guidelines' goal is to improve the quality of the preparation for animal studies. It covers three areas: Formulation of the study, Dialogue between scientists and the animal facility, and Methods. Each area consists of two parts – general and the part for fish researchers. Each of the three broad areas is further divided into subcategories. The PREPARE guidelines are especially beneficial for the research that involves work with fish.

The formulation of the study consists of the subcategories: Literature search, Legal issues, Ethical issues, harm-benefit assessment, and humane endpoints, and Experimental design and statistical analysis.

The dialogue between scientists and the animal facility area is further divided into: Objectives and timescale, funding, and division of labor, Facility evaluation, Education and training, and Health risks, waste disposal, and decontamination.

The methods area is subdivided into the following parts: Test substances and procedures, Experimental animals, Quarantine and health monitoring, Housing and husbandry, Experimental procedures, Humane killing, release, reuse, or rehoming, and Necropsy.

Non-Technical Summary

As a part of the application to work with animals, a researcher writes a non-technical summary (NTS) [9]. The NTS is an integral part of the application and the only part visible to the general public. Namely, the NTS helps lay people understand why animals are used in scientific research and how the researcher will implement the replacement, reduction, and refinement (3Rs) policy. The non-technical project summary is anonymous and does not contain identifying or other sensitive information such as intellectual property. The NTS must include the objective and description of the envisaged work, the number of used animals, the predicted harms to animals, and the expected benefit of the planned research.

In the NTS, the researcher must prove that there is no suitable alternative for the use of animals; that the benefits of the project outweigh the effects on the animals; that the number of animals involved is minimized, and that animals are treated in a humane manner and following the best standards of modern animal husbandry with as little suffering as possible. The protected animals range from any living vertebrates (e.g., nonhumane primates, large and small animals) to any living cephalopod (e.g., octopus, squid, cuttlefish, and chambered nautilus) [10,11].

The NTS must be written in plain, non-biomedical language without abbreviations and easily understandable to lay person. The applicant also must explain any technical terms. The NTS should be concise, of publishable quality, with recommended 500 words (one A4 page), and able to stand alone. For consistency purposes, a working document on non-technical project summaries contains Annex I, Template/Headings for a Non-Technical Summary, and Annex II, an illustrative example of a completed Non-Technical Summary [9,10,11].

When the application is approved, the NTS is published on the Member State/National Contact Point webpage and accessible for at least five years after completing a project. However, suppose the application undergoes a retrospective assessment. In that case, the NTS is updated with these new results and is accessible for at least five years after completing the retrospective assessment.

The ethics review process

The ethics review follows the scientific review. The ethical basis for the proper justification of the proposal is previous EC scientific evaluations. Only proposals envisaged for funding or ones that are very close to the cutoff line will undergo ethics evaluation. The EC officers/(external) experts/reviewers do not additionally judge the proposal's science. The EC ethics officers that work full time on ethics and independent ethics expert will conduct proposals' ethics evaluation. Moreover, ethics officers are in contact with the applicants to help them address the ethics issues before and after the signature of the grant.

The ethics officers will conduct a pre-screening step and clear proposals where the ethics issues are addressed well. The rest of the proposals that did not adequately address the ethical issues, or involve particularly sensitive ones, must undergo an ethics screening process conducted by independent experts.

Before the ethics screening starts, selected independent experts must confirm that they have no conflict of interest with the assigned proposals and their corresponding applicants and commit to confidentiality.

The ethics screening phase is a few-step procedure that slightly varies among EC agencies. Namely, it can be conducted entirely remotely or with the panel meeting in Brussels. In the first step, each proposal is reviewed remotely by 2 or 3 experts (depending on the agencies) via the EC evaluation system. The ethics expert reads the entire proposal, including all annexes, but does not judge science and assesses the ethical component only. Each reviewer writes an ethics individual evaluation report for each screened proposal during this step. These individual reports are merged into a draft consensus report that serves for discussion during panel meetings.

Depending on the number of proposals, each panel consists of around ten ethics experts covering all ethical fields. During the panel meeting, independent experts, with the help of ethics officers, produce an ethics summary (consensus) report for each proposal. Some summary reports contain ethics recommendations without requirements that are not contractual obligations. On the other hand, some summary reports have ethics requirements that must be clarified or addressed. Ethics officers send ethics summary reports to applicants and, based on their replies, clear some proposals while others forward to the next step – the assessment panel. Usually, the projects that raise sensitive ethics issues or for which information was lacking are forwarded to the assessment panel. For these sensitive proposals and proposals lacking essential information, the ethics review panel and ethics assessment panel can request an independent ethics officer or ethics advisory board to be on the project to submit ethics reports as proposal deliverables periodically.

In the ethics assessment panel, which is also multidisciplinary, ethics experts perform a more in-depth ethics analysis of the proposal. At the end of the ethics review process, outstanding ethics requirements become contractual obligations as part of the grant agreement. The proposal is then assigned to an ethics officer responsible for the ethics monitoring of these requirements.

The assessment panel can also designate some proposals for ethics check/ audit after the start of the project. There are cases when an assessment panel can only make decisions with ethical implications during the project's lifetime or when a closer ethics oversight is needed. The ethics check is carried out by the European Commission or the relevant funding body and with the support of the ethics experts. The ethics check activities may require additional information, a principal investigator interview in Brussels, or an on-site audit.

The animal work review process

The animal review process is governed by the EU Directive 2010/63/EU, Document 32021R0695, Version 26/06/2019 [12]. The work with genetically modified organisms is additionally regulated by the EU Directive 2001/18/ EC, Document 02001L0018-20210327, Version: 27/03/2021 [13].
The ethics reviewers are focused on the following issues:

• Do planned activities involve work with animals, and are they vertebrates;

- Do planned activities involve work with nonhuman primates;
- Do planned activities involve work with wild type or genetically modified animals;
- Do planned activities involve work with cloned farm animals;
- Do planned activities involve work with endangered species;
- What animal species will be involved;
- Did the applicant enclose a statistical power analysis with the estimated number of animals needed for the experiments?

It is expected from applicants to enclose some of the following documents related to their works:

- A copy of the facility authorization(s) where animal experiments will be conducted;
- Copies of the application(s) for project authorization(s) to the competent authorities and the final approval(s);
- The applicant must describe the nature and severity of the unavoidable harm that will cause to the animals and the measures to minimize the harm and enhance the welfare of the animals. The applicant must also provide details on the procedural techniques, anesthesia, analgesia protocols, post-procedural follow-up and care, other refinement methods, and the procedure-specific early identification of pain, humane endpoints, and euthanasia methods.
- The applicant must provide copies of the training certificates of staff involved in animal experiments;
- The applicant must provide copies of training certificate(s)/personal license(s) that certify the competence of the responsible person in charge of designing the projects and procedures of the animal experiments;
- In case nonhuman primates are involved in the project, the applicant must justify why the research aim cannot be achieved using species other than nonhuman primates;
- The applicant must provide a statement with their commitment to keeping the personal history files of the nonhuman primates involved in the project for at least five years after the end of the project. The applicant must provide copies of these files upon request.

However, potential ethical issues related to the use of animals may not be limited exclusively to working with animals. Namely, suppose work with animals is envisaged in low and/or lower-middle income countries. In that case, the applicant must develop a plan for benefit sharing or capacity building in these countries [1,4,14]. Also, suppose the work is planned in a non-EU country. In that case, the researcher must address the issues of using local resources (animal tissues samples, live animals, endangered animals) and the export/import to or from the EU (a material transfer agreement is needed). An additional issue related to the researchers working in non-EU countries is if their activities could put them at risk.

Working with genetically modified animals or genetically modifying them brings concern to Environment, Health, and Safety [15]. The issue is if methods, materials, or experiments could harm the environment, animals, or research staff. There is a set of regulations related to environmental protection and safety. Namely:

- Directive 2000/54/EC on the protection of workers from risks related to exposure to biological agents at work [16];
- Directive 2001/18/EC on the deliberate release into the environment of genetically modified organisms [13];
- Directive 2009/41/EC [17] and 98/81/EC [18] on the contained use of genetically modified micro-organisms;
- Regulation EC No 1946/2003 on transboundary movements of genetically modified organisms [19];
- Directive 2008/56/EC on marine environmental policy [20];
- Council Directive 92/43/EEC on conservation of natural habitats and of wild fauna and flora [21];
- and Council Regulation EC No 338/97 on the protection of species of wild fauna and flora [22].

Moreover, some specific and potentially newly arising issues, such as artificial intelligence and animal work or animal-machine integration, xenobots, synthetic biology, chimeras or germ cell organoids could be addressed in the section 'Other Ethics Issues'.

The specific purpose of the ethics review of the proposed research is to protect the welfare of animals and the researchers' safety. Namely, the work with animals may jeopardize the safety or security of researchers or lead to a lack of capacity to enforce the relevant ethical standards and guidelines that may affect researchers and/or animals.

References

1. EUR-Lex. Document 32021R0695
   Regulation (EU) 2021/695 of the European Parliament and of the Council of 28 April 2021 establishing Horizon Europe – the Framework Programme for Research and Innovation, laying down its rules for participation and dissemination, and repealing Regulations (EU) No 1290/2013 and (EU) No 1291/2013 (Text with EEA relevance)
   https://eur-lex.europa.eu/legal-content/EN/ALL/?uri=CELEX:32021R0695
2. European Commission: EU Grants - How to complete your ethics self-assessment. Version 2.0. 13 July 2021
   https://ec.europa.eu/info/funding-tenders/opportunities/docs/2021-2027/common/guidance/how-to-complete-your-ethics-self-assessment_en.pdf
3. AllEA: The European Code of Conduct for Research Integrity.
   https://allea.org/code-of-conduct/
4. The Global Code of Conduct for Research in Resource-Poor Settings
   https://www.globalcodeofconduct.org/wp-content/uploads/2018/05/Global-Code-of-Conduct-Brochure.pdf
5. The ARRIVE guidelines 2.0. *https://arriveguidelines.org/arrive-guidelines*
6. Percie du Sert N, Ahluwalia A, Alam S, Avey MT, Baker M, Browne WJ, Clark A, Cuthill IC, Dirnagl U, Emerson M, Garner P, Holgate ST, Howells DW, Karp NA, Lazic SE, Lidster K, MacCallum CJ, Macleod M, Pearl EJ, Petersen OH, Rawle F, Reynolds P, Rooney K, Sena ES, Silberberg SD, Steckler T, Würbel H. The ARRIVE guidelines 2.0: Updated guidelines for reporting animal research. Pbio July 14, 2020. *https://doi.org/10.1371/journal.pbio.3000410*
7. Percie du Sert N, Hurst V, Ahluwalia A, Alam S, Avey MT, Baker M, Browne WJ, Clark A, Cuthill IC, Dirnagl U, Emerson M, Garner P, Holgate ST, Howells DW, Hurst V, Karp NA, Lazic SE, Lidster K, MacCallum CJ, Macleod M, Pearl EJ, Petersen OH, Rawle F, Reynolds P, Rooney K, Sena ES, Silberberg SD, Steckler T, Würbel H. Reporting animal research: Explanation and elaboration for the ARRIVE guidelines 2.0. Pbio July 14, 2020. *https://doi.org/10.1371/journal.pbio.3000411*
8. The PREPARE guidelines. *https://norecopa.no/prepare*
9. Working document on Non-Technical Project summaries by National Competent Authorities for the implementation of Directive 2010/63/

EU on the protection of animals used for scientific purposes. *https://ec.europa.eu/environment/chemicals/lab_animals/pdf/Recommendations%20 for%20NTS.pdf*

10. Understanding Animal Research. Bella Williams: Guide to writing non-technical summaries. Posted on 23/07/18.
    https://www.understandinganimalresearch.org.uk/news/guidance-for-writing-a-nts

11. Taylor, K., Rego, L. and Weber, T. (2018) "Recommendations to improve the EU non-technical summaries of animal experiments", *ALTEX - Alternatives to animal experimentation*, 35(2), pp. 193–210. doi: 10.14573/altex.1708111.

12. EUR-Lex. Document 02010L0063-20190626
    Consolidated text: Directive 2010/63/EU of the European Parliament and of the Council of 22 September 2010 on the protection of animals used for scientific purposes (Text with EEA relevance)Text with EEA relevance. Version 26/06/2019 *https://eur-lex.europa.eu/legal-content/EN/ TXT/?uri=CELEX%3A02010L0063-20190626*

13. Eur-Lex. Document 02001L0018-20210327
    Consolidated text: Directive 2001/18/EC of the European Parliament and of the Council of 12 March 2001 on the deliberate release into the environment of genetically modified organisms and repealing Council Directive 90/220/EEC. Version: 27/03/2021 *https://eur-lex.europa.eu/legal-content/EN/ TXT/?uri=CELEX%3A02001L0018-20210327*

14. Horizon Europe (HORIZON) Programme Guide Version 1.5 01 February 2022 *https://ec.europa.eu/info/funding-tenders/opportunities/docs/2021-2027/horizon/ guidance/programme-guide_horizon_en.pdf*

15. Trbovich AM. Genetically modified organisms and bioethics. In Todorovic Z, Prostran M, Turza K. Bioethics and Pharmacology: Ethics in Preclinical and Clinical Drug Development. Transworld Research Network, Kerala, India; 2012, 59-72.

16. EUR-Lex. Document 02000L0054-20200624
    Consolidated text: Directive 2000/54/EC of the European Parliament and of the Council of 18 September 2000 on the protection of workers from risks related to exposure to biological agents at work (seventh individual directive within the meaning of Article 16(1) of Directive 89/391/EEC). Version 24/06/2020
    https://eur-lex.europa.eu/legal-content/EN/TXT/?uri=CELEX:02000L0054-20200624

17. EUR-Lex. Document 32009L0041
    Directive 2009/41/EC of the European Parliament and of the Council of 6 May 2009 on the contained use of genetically modified micro-organisms (Recast) (Text with EEA relevance) *https://eur-lex.europa.eu/legal-content/EN/ TXT/?uri=CELEX:32009L0041*

18. EUR-Lex. Document 01998L0081-20090610
    Consolidated text: Council Directive 98/81/EC of 26 October 1998 amending Directive 90/219/EEC on the contained use of genetically modified micro-

organisms.   Version   10/06/2009   *https://eur-lex.europa.eu/legal-content/EN/TXT/?uri=CELEX%3A01998L0081-20090610*

19. EUR-Lex. Document 32003R1946

Regulation (EC) No 1946/2003 of the European Parliament and of the Council of 15 July 2003 on transboundary movements of genetically modified organisms (Text with EEA relevance)

https://eur-lex.europa.eu/legal-content/en/TXT/?uri=CELEX:32003R1946

20. EUR-Lex. Document 02008L0056-20170607

Consolidated text: Directive 2008/56/EC of the European Parliament and of the Council of 17 June 2008 establishing a framework for community action in the field of marine environmental policy (Marine Strategy Framework Directive) (Text with EEA relevance). Version 07/06/2017

https://eur-lex.europa.eu/legal-content/EN/TXT/?uri=CELEX%3A02008L0056-20170607

21. EUR-Lex. Document 01992L0043-20130701

Consolidated text: Council Directive 92/43/EEC of 21 May 1992 on the conservation of natural habitats and of wild fauna and flora. Version 01/07/2013

https://eur-lex.europa.eu/legal-content/EN/TXT/?uri=CELEX%3A01992L0043-20130701

22. EUR-Lex. Document 31997R0338

Council Regulation (EC) No 338/97 of 9 December 1996 on the protection of species of wild fauna and flora by regulating trade therein. Version 19/01/2022

https://eur-lex.europa.eu/legal-content/EN/ALL/?uri=CELEX%3A31997R0338

BIOETHICS STANDARDS IN GOOD LABORATORY PRACTICE: THE CURRENT FOLLOW-UP

Miroslav Radenković

Department of Pharmacology, Clinical Pharmacology and Toxicology, Faculty of Medicine University of Belgrade, Serbia

Abstract

The current state of scientific development allows for biomedical laboratory experiments, that use various physical, chemical, or biological test systems. Moreover, pre-clinical research experiments represent one of the crucial links in understanding pharmacological and other properties of the future drug. Still, the highest level of investigation can be supplied only if there is a globally recognized framework for conducting experimental investigations that adheres to the strictest bioethical norms. The OECD Principles of Good Laboratory Practice were first established in 1981 and revised in 1997 with the goal of creating a recognized system of quality assurance for test results used in evaluating chemicals for their impacts on human health and the environment. This quality system addresses the administrative process itself, as well as the settings in which laboratory experiments are planned, carried out, monitored, documented, and reported. From a bioethical standpoint, various strategies are being developed or are already in use to limit, refine, or replace the use of animals in pre-clinical research. Moreover, significant efforts are undertaken on a regular basis to bring scientists organized to enhance the implementation of ethical standards anytime animals are used, as well as to push for more trust, transparency, and engagement with the public.

Keywords: good laboratory practice, bioethics, animal protection

Introduction

The current level of scientific progress allows biomedical studies to use various physical, chemical, or biological test systems, such as numerous animal models that can actually mimic different human pathological conditions and are sufficiently similar to humans in relevant aspects of the biological phenomenon or disease being studied [1]. The fundamental idea, especially in medical research involving animal models, is that they can genuinely provide further knowledge about the condition under investigation as well as better diagnostic, therapeutic, and preventative options. As a result, as recently said, biomedical research encompasses a broad range of activities ranging from the study of fundamental physiological processes to the understanding of disease causes and the creation of treatment options [2].

We know that by definition animal experimentation refers to any procedure that alters an animal's well-being and has the potential to cause pain, suffering, anguish, or discomfort; and that the goal of the procedure is always to reveal biological phenomena, even if the results are not completely compatible with human beings [3]. Therefore, it's important to remember that animal in vitro and in vivo models can only provide a rough estimate of a specific human homeostasis disturbance and can never entirely replicate a pathological core of illness. Extrapolations from animal research can be made in a meaningful fashion, but only as long as the scientific goals are appropriate and ethically supported.

According to previous facts, it is always necessary to make an initial assessment of a reasonable and acceptable bioethical framework, as well as a harm-to-benefit analysis for using experimental animals, especially when there is a serious concern that the knowledge gained will be minor and insignificant. Furthermore, quality assurance procedures should be essential and implemented at all times during the process of providing trustworthy and reproducible data. As a result, the goal of a quality assurance program would be to ensure data consistency, traceability, and repeatability [4]. Considering the preceding considerations, it is reasonable to assume that the highest quality of investigation process can be provided only if there is a solid and large-scale recognized system for conducting experimental investigations involving various test systems, which would also adhere to the highest bioethical standards. In other words, regardless of the goal of the study or where it is conducted, namely academic, industry, or a contract laboratory, rigorous and thorough application of core scientific practices is required for safety choices [5].

The OECD Principles of Good Laboratory Practice

The Principles of Good Laboratory Practice (GLP) of the Organisation for Economic Co-operation and Development (OECD) were developed by an Expert Group on GLP established in 1978 under the Special Programme on Chemicals Control. The Expert Group's work was based on the GLP regulations, which were issued by the US Food and Drug Administration in 1976 and were approved as international standards for non-clinical laboratory investigations [6]. The OECD Principles of Good Laboratory Practice were implemented soon after, in 1981. Specifically, they were originally included in Annex II of the OECD Council Decision on Mutual Acceptance of Data [C(81) 30 (Final)] [7]. On November 26, 1997, the final revisions to the GLP principles were adopted [C(97) 186 (Final)] [8]. The main goal was to create a globally recognized quality assurance system for test results that would be utilized in evaluating chemicals for their impacts on human health and the environment. This quality system would cover the administrative process as well as the settings under which laboratory experiments are planned, carried out, monitored, documented, and reported.

The importance of the GLP principles clearly lies in the ongoing promotion of a concept of mutual acceptance of data, thereby avoiding possible duplication of scientific efforts and duplicative animal testing, thereby reducing animal suffering, and thus leading to significant progress in resolving important animal welfare issues [9]. As a result, if independent laboratories can rely on experimental results acquired by peers in other nations, a significant amount of time and important resources can be saved. This would also help to close the technological gap, as well as improve human health and environmental safety.

The OECD GLP standards must be followed to ensure data quality, integrity, and repeatability. Furthermore, only complete implementation of scientific validity and data quality requirements can ensure reliable public health and environmental protection. Only proper synchronization and harmonization amongst the primary stakeholders of the OECD GLP principles, namely sponsors, regulatory bodies, monitors, and test facilities, can achieve this [10, 11].

The pre-clinical investigation of natural or synthetic chemicals that have been preliminary screened and further developed for potential pharmaceutical, pesticide, or cosmetic products, as well as veterinary drugs, food or feed additives, and industrial chemicals, should follow the OECD GLP principles. Furthermore, these criteria are required by rules for the reason of registering or licensing the above-mentioned products, unless expressly disallowed by national legislation.

The Organisation for Economic Co-operation and Development (OECD) principles of GLP are shortly overviewed in the following sections [6, 12].

It is highlighted that the management of each test facility should ensure that GLP principles are followed. A sufficient number of qualified individuals, as well as relevant facilities, equipment, and supplies, should be available in this manner. Each member of the research team should have a clear understanding of their responsibilities, which means that relevant Standard Operating Procedures should be readily available and followed at all times. It is necessary to guarantee that test facility supplies meet the standards for their use in a study, as well as that test and reference items are properly identified. The resulting raw data must be adequately documented and recorded. With designated employees participating, a suitable Quality Assurance Program must be proposed. The study plan, final report, raw data, and supporting material must all be archived properly after the study has been completed.

Apart from the responsibility to record raw data promptly and accurately, one of the most important tasks for study personnel is to follow the principle that any deviation from the study plan or already approved Standard Operating Procedures should be documented and communicated directly to the study director and/or, if appropriate, the principal investigator (s). From a bioethical standpoint, study staff should take health precautions to reduce risk(s) to themselves and to ensure the study's integrity.

To ensure complete compliance with GLP principles, the test facility should have a documented Quality Assurance Program, which should be carried out by individuals assigned by and directly responsible to management, who are familiar with the test procedures but are not directly involved in the study's conduct. Inspections can be classified as study-, facility-, or process-based, according to approved Standard Operating Procedures. As a result, it must be determined whether all studies are conducted in accordance with the adopted GLP Principles, whether study plans and Standard Operating Procedures have been made available, and whether all methods, procedures, and observations are accurately and completely described in the final report, as well as whether the reported results accurately and completely reflect the raw data of the studies.

The GLP Principles include specific suggestions and guidelines for the size, structure, and location of the test facility. They should all be suitable to match the study's requirements while also allowing for the elimination of any disturbance that could jeopardize the investigation's authenticity. Furthermore, the different activities in the test facility must be separated to a sufficient degree. As a result, the test facility should contain enough rooms or sections to ensure that distinct test system operations are isolated. One of the most important activities would be to maintain the identity, concentration, purity, and stability of test and reference chemicals. It should be simple to store and retrieve study plans, raw data, final reports, and kept samples in a secure manner. The test facility's equipment should be suitable for its intended use and should be inspected, cleaned, maintained, and calibrated on a regular basis. The test system (physical, chemical, and biological - animals, plants, and so on) should be unaffected by the apparatus and materials utilized. All chemicals, reagents, and solutions used must be properly labeled and other information such as the manufacturer's name, preparation date, and stability must be readily available.

Written Standard Operating Procedures should be one of the most important documents in any testing facility. This is an official document that must be approved by test facility management in order to ensure the quality and integrity of data generated by that test facility. Additional modifications, on the other hand, should always be approved by test facility management. All members of the study team should have immediate access to Standard Operating Procedures relevant to the activities being undertaken. Detailed information about test/reference items, equipment-related issues, materials, reagents, record keeping-reporting-storage-retrieval, test system(s), or quality assurance methods can be found in these files.

An investigation should be carried out in accordance with the study design, financial resources available, and the time frame initially set. All research entries should be recorded directly, accurately, and quickly. A properly designed final report should be created at the conclusion of each study, which is usually signed by the study director. The final report should include all relevant study data, including those related to the quality assurance program, a description of the methods and materials used, the obtained results with adequate discussion and conclusions, as well as storage information about the study plan, test and reference item samples, specimens, raw data, and the final report.

Several consensus documents covering the quality assurance unit, guides for compliance monitoring procedures for GLP, application of the GLP principles to field studies, application to in vitro studies, establishment and control of archives, and application of the GLP principles to short studies were developed for the OECD GLP principles in order to provide additional characterization and precision to originally established principles [13]. However, it should be emphasized that knowledge of and confidence in national good laboratory practice compliance monitoring programs, which ensures the acceptability for regulatory purposes of non-clinical environment and health safety data on chemicals and chemical products tested in OECD countries, represents the cornerstone of mutual data acceptance [14].

Bioethical Standards

The utilization of animals for the benefit of humans is a difficult issue in many places throughout the world. Moral and ethical difficulties involving the use of animals in scientific research, education, and testing are complicated by cultural morality and traditions, the role of animals in many religions, individual and cultural ethical standards, and the diversity of concern for certain species [15]. The fact that scientists who use animals in research must justify the number of animals to be used, as well as methodology to be used, and ethics committees that are required to review these proposals for experiments to be undertaken on animals, all represent the common ground for global standardization and harmonization of any scientific work [16]. Many scientific publications now include ethical clearances as a standard, as well as compulsory feature [17]. The number of animals to be utilized may usually be predicted simply from previous experience, especially when using well-established methodologies and protocols [18]. Furthermore, if a null-procedure protocol is conceivable, implying that the test subject or isolated tissue/organ can be utilized in a traditional before/after experiment, the number of laboratory animals used will be significantly lower [19, 20]. As a result, no matter how elaborate an experimental design is proposed, or how complex and significant the working hypothesis appears to be, one should always try to employ the smallest number of animals necessary to accomplish an effective data analysis and maybe attain observable statistical significance. That is why biomedical researchers are recommended to engage a statistician early in the design process, so that they may efficiently carry out their experiments and evaluate them, so that they can extract all valuable information from the generated records.

It is critical to utilize animals correctly, which includes avoiding or minimizing discomfort, distress, and pain when doing so is consistent with acceptable scientific standards, and moreover, investigators should assume that techniques that cause pain or distress in humans may also cause pain or distress in other animals unless the contrary is proven [15]. In light of the foregoing, it's comprehensible that a variety of in vitro and equivalent ex vivo approaches are being developed or are already in use to reduce, refine, or replace (The 3Rs) animal use in pre-clinical research, as defined by Russell and Burch (1959) [21]. As a result, the reduction options would include strategies for acquiring equivalent information with fewer animals or obtaining more information with a given number of animals [22]. The refinement options aim to eliminate or minimize pain and distress or improve animal well-being, with much pain and distress being reduced or eliminated with proper use of anesthetics and analgesics, and well-being being improved by proper handling, using appropriately sized cages, and group housing of social species [23]. Finally, the alternative possibilities would include particular strategies that would allow a given aim to be achieved without the use of animals in experiments or other scientific procedures [22]. This could include in vitro experiments on isolated organs or primary cell cultures obtained from a properly sacrificed animal [18, 24], or in vitro experiments using cells directly obtained from the living organism. Unfortunately, the interpretation of the 3Rs standards for some researchers may be influenced by personal morality, the recourse account for compliance, or even the cost of pricey in vivo animal–replacement technologies. [3].

The essential and accompanying changes are also being implemented in present guidelines, standards and related actions to allow the favorable reception of data by employing completely new approaches. As a result, the most recent scientific developments in non-animal methodologies would encompass not only a wide range of cell and tissue culture systems, but also a wide range of high-throughput and in silico screening approaches. As a result, testing laboratories must follow acceptable scientific procedures and use calibration and standardization methodologies that are appropriate for the components of their assay system [25]. In accordance, user surveys and post-implementation software performance evaluations may be required to determine whether or not specific laboratory software is fulfilling its functions as stated [26].

Even though many of the most difficult concerns concerning the use of animals in research remain unresolved, philosophical theory and moral reasoning have contributed significantly to bringing the debate ahead in a reasonable manner [27]. Significant efforts are made on a regular basis to bring the scientific community together to advance the implementation of ethical principles such as the 3Rs whenever animals are used in research, as well as to call for more trust, transparency, and communication on the sensitive topic of animals in research [2]. To continue, the responsible treatment of animals in science through advanced voluntary accreditation programs should be strongly supported, where for example, the

Association for Assessment and Accreditation of Laboratory Animal Care International (AAALAC International) represents a fine model of excellence in this susceptible area [28]. As a result, in order to improve widespread knowledge of the significance and usefulness of conducting animal studies with the goal of improving human health, the general public should have access to variety resources of information. Consequently, if community dialogue is more trustworthy, the science and the animal welfare should not be in conflict. This would be an outstanding method to raise awareness and gain a better grasp of the reality that animal studies are unquestionably important for biomedical advancement today and in the future. This would also greatly boost the chances of establishing a genuine and functioning link between basic and applied research.

In short, from a bioethics perspective, high standards of experimental investigation need to be established so as to ensure ethical applicability, consistent safety, and a high sense of responsibility, so long as this is realistic. As such, the future development of consultation and training related to further recognition and improvement of the GLP guidelines is expected to be an ongoing process involving new technologies and innovative learning methods, especially Internet-based learning techniques [29]. As a result, continual bioethics education is more than important, particularly in biological fields–related universities [30].

Author's statement

The opinions and the analysis expressed in this manuscript are those of the author and do not necessarily represent the viewpoints of the referenced authorities.

References

1. Perry P, Joynson C. The ethics of animal research: A UK perspective. ILAR J 2007;48(1):42-46.
2. Treue, S., and Hengartner, M. Basel Declaration Annual Report 2010–2011. S. Treue and M. Hengartner (Eds.), Basel Declaration, 2011, Zurich.
3. Aguilar TAF, Guarneros Bañuelos E. Bioethics in the Use of Experimental Animals. In: Morales-González JA, Nájera MEA, Eds. Reflections on

Bioethics [Internet]. London: IntechOpen; 2018. Available from: https://www.intechopen.com/chapters/59654 doi: 10.5772/intechopen.74520.

4.   Hartung T, Balls M, Bardouille C, Blanck O, Coecke S, Gstraunthaler G, et al. Good cell culture practice: ECVAM good cell culture practice task force report 1. ATLA Altern Lab Anim 2002;30(4):407-414.

5.   Becker RA, Janus ER, White RD, Kruszewski FH, Brackett RE. Good laboratory practices: Becker et al. Respond. Environ Health Perspect 2010;118(5):A194-A195.

6.   Akyar I. GLP: Good Laboratory Practice. In: Eldin, A. B. , Ed. Modern Approaches to Quality Control [Internet]. London: IntechOpen; 2011. Available from: https://www.intechopen.com/chapters/22127 doi: 10.5772/19823.

7.   Organisation for Economic Co-operation and Development. 1981, Decision of the Council concerning the Mutual Acceptance of Data in the Assessment of Chemicals. [C(81)30].

8.   Organisation for Economic Co-operation and Development. 1998, Principles on good laboratory practice (as revised in 1997). OECD, Paris (Series on principles of GLP and compliance monitoring No.1, ENV/MC/CHEM(98)17).

9.   Koëter HBWM. Mutual acceptance of data: Harmonised test methods and quality assurance of data - The process explained. Toxicol Lett 2003;140-141:11-20.

10. Hendriks R, Van Den Eynde H, Coussement W. Responsibilities of test facility management and sponsor in a GLP environment. Ann Ist Super Sanita 2008;44(4):407-408.

11. Beernaert H, Vanherle A-, Bertr S. Critical aspects in implementing the OECD monograph No. 14 "The application of the principles of GLP to in vitro studies". Ann Ist Super Sanita 2008;44(4):348-356.

12. Radenković M. The OECD principles of good laboratory practice and the current bioethics, In: Bioethics and Pharmacology: Ethics in Preclinical and Clinical Drug Development; Todorović Z, Prostran M and Turza K, Eds., Transworld Research Network, Kerala, India, 2012: 43-50.

13. Huntsinger DW. OECD and USA GLP applications. Ann Ist Super Sanita 2008;44(4):403-406.

14. Turnheim D. Current state of the implementation of the OECD GLP principles in the OECD member countries and non-member economies in light of the outcome of the 1998-2002 pilot project of mutual joint visits. Ann Ist Super Sanita 2008;44(4):327-330.

15. Simmonds RC. Bioethics and Animal Use in Programs of Research, Teaching, and Testing. In: Weichbrod RH, Thompson GAH, Norton JN, Eds. Management of Animal Care and Use Programs in Research, Education, and Testing. 2nd edition. Boca Raton (FL): CRC Press/Taylor & Francis; 2018. Chapter 4. doi: 10.1201/9781315152189-4

16. Dell RB, Holleran S, Ramakrishnan R. Sample size determination. ILAR J 2002;43(4):207-212.

17. Radenković M, Stojanović M, Topalović M. Contribution of thromboxane A2 in rat common carotid artery response to serotonin. Sci Pharm 2010;78(3):435-443.

18. Radenković M, Stojanović M, Janković R, Topalović M, Stojiljković M. Combined contribution of endothelial relaxing autacoides in the rat femoral artery response to CPCA: An adenosine A2 receptor agonist. Sci World J 2012;2012.

19. Grbović L, Radenković M, Prostran M, Pešić S. Characterization of adenosine action in isolated rat renal artery: Possible role of adenosine A2A receptors. Gen Pharmacol : Vasc Syst 2000;35(1):29-36.

20. Radenković M, Grbović L, Pešić S, Stojić D. Isolated rat inferior mesenteric artery response to adenosine: Possible participation of Na + /K + -ATPase and potassium channels. Pharmacol Rep 2005;57(6):824-832.

21. Russell WMS and Burch RL. The Principles of Humane Experimental Technique. Methuen & Co. LTD., 1959, London. Available at: https://caat. jhsph.edu/principles/the-principles-of-humane-experimental-technique.

22. Zurlo J, Rodacille D, Goldberg AM. The three Rs: The way forward. Environ Health Perspect 1996;104(8):878-880.

23. Goldberg AM, Zurlo J, Rudacille D. The three Rs and biomedical research. Science 1996;272(5267):1403.

24. Zebedin E, Koenig X, Radenkovic M, Pankevych H, Todt H, Freissmuth M, et al. Effects of duramycin on cardiac voltage-gated ion channels. Naunyn-Schmiedeberg's Arch Pharmacol 2008;377(1):87-100.

25. Gupta K, Rispin A, Stitzel K, Coecke S, Harbell J. Ensuring quality of in vitro alternative test methods: Issues and answers. Regul Toxicol Pharmacol 2005;43(3):219-224.

26. Whartenby GJ, Robinson PL, Weinberg S. The Good Automated Laboratory Practices. Drugs Pharmac Sci, 2007, 168: 131-149.

27. Russow L. Bioethics, animal research, and ethical theory. ILAR J 1999;40(1):15-21.

28. Gettayacamin M, Retnam L. AAALAC International Standards and Accreditation Process. Toxicological Research. 2017;33(3):183–9.

29. Long D. Developments in consultancy and training in the GLP arena: 1980 to 2020. A personal view. Ann Ist Super Sanita 2008;44(4):409-416.

30. Dos Santos Feijó AG, Crippa A, Steffen JL. Proposal for bioethics education in Animal Ethics. Rev. Bioetica & Derecho, 2016, 36: 85-91.

BIOETHICS AND PHARMACOLOGY: THE PRECLINICAL DRUG DEVELOPMENT

Zoran Todorović

University of Belgrade School of Medicine and University Medical Center "Bežanijska kosa", Belgrade, Serbia

Abstract

Bioethics is an inseparable part of drug development, i.e., contemporary pharmacology and toxicology. A holistic approach to bioethics applies to all forms of life and the environment and is the legacy of the founders of bioethics, Jahr and Potter. Compliance with modern legislation related to the welfare of experimental animals (e.g., Directive 2010/63/EU and similar regulations in the UK, USA, and other countries) is permeated by 3Rs (Russell and Birch 1959), as well as the principles of animal bioethics set by Dr. Marshall Hall. Preclinical drug development consists of several phases and involves *in vivo* experiments. A particular challenge is testing drugs affecting cognition, biological drugs, and experiments involving primates. Also, animal bioethics is essential for preclinical pharmacokinetics and toxicology. All these elements are considered by ethical committees that analyze both the ethical and scientific sides of the experiment. In decision-making, they start from the utilitarian model of estimating the experiment's benefits, risks, and quality (the Bateson decision-making cube).

Keywords: bioethics, neuroethics, pharmacology, toxicology, drug development

Introduction

In experimental biomedicine, pharmacology, a science of drugs established in the mid-nineteenth century, has a special place. The first pharmacological institute was founded at the University of Dorpat (today Tartu in Estonia), and the founder was the German pharmacologist

Rudolph Bucheim (1820-1879). Bucheim reshaped pharmacology as an exact experimental science instead of an empirical discipline. In particular, he introduced bioassay in drug development and research, an analytical method of a qualitative and quantitative assessment of the presence and potency of chemical substances in living tissues that was later promoted by the British Nobel Prize laureate Sir John Vane (1927-2004)1.

Pharmacology has evolved in two ways. First, the study of the effect of drugs on living systems took place in the milieu of experimental laboratories. Second, there was a need to test new drugs in clinical settings, and much later, clinical trials developed. Drugs could be defined as chemical or biological substances that act on living systems, excluding nutrients or food ingredients in their composition. Methods used in pharmacology cover the broadest possible range, from isolated ion channels (patch-clamp technique), receptors, genes, metabolites, and signaling molecules, through isolated organs and whole animals to humans and the entire population. Animal experiments are an essential part of core pharmacology disciplines such as pharmacokinetics, pharmacodynamics, and toxicology, as well as the border fields - e.g., pharmacogenetics and pharmacogenomics. If the drug development continuum, i.e., the period from its discovery to withdrawal from the market, is divided into phases, the first is preclinical development, which includes experiments on animals, their organs, tissues, and cells (Fig. 1).

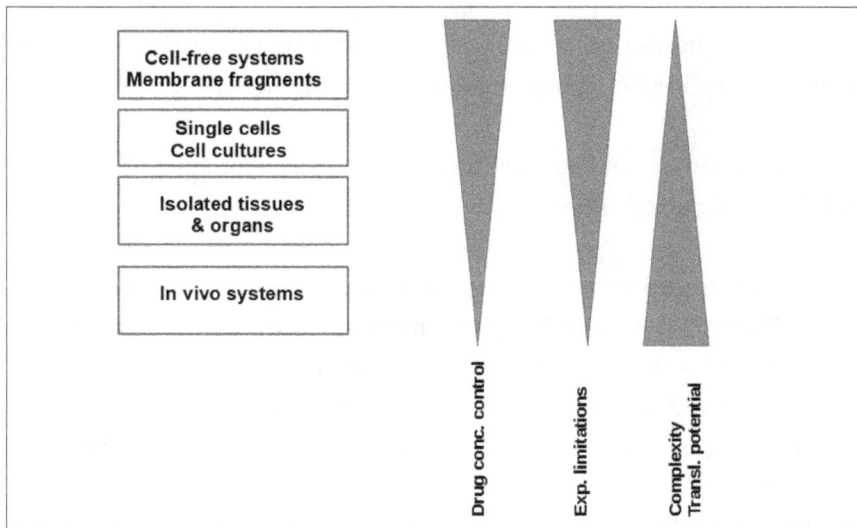

Fig. 1 Drug response systems

The starting point in assessing the ethics of animal experiments are the Principles of Research in Physiology (Dr. Marshall Hall, 1790-1857):

- We should never have recourse to experiment, in cases in which observation can afford us the information required.
- No experiment should be performed without a distinct and definite object, and without the persuasion, after the maturest consideration, that that object will be attained by that experiment, in the form of a real and uncomplicated result.
- We should not needlessly repeat experiments which have already been performed by physiologists of reputation.
- A given experiment should be instituted with the least possible infliction of suffering...
- Physiological experiments should, if possible, only be instituted with the aid of competent witnesses (the least possible number of animals).

Animal experimentation refers to the scientific study of animals, usually in a laboratory, for the purpose of gaining new biological knowledge or solving specific medical, veterinary medical, dental, or biological problems (2). At the same time, in any case, the welfare of experimental animals is violated, but this is done to the least possible extent and in order to acquire new knowledge.

Concerning the use of animals for experimental purposes, the starting point is the concept of animal welfare and the 3R rule (3Rs).

Animal welfare is not based on how people feel about animals, although that is not neglected either.

- The British utilitarian ethicist Jeremy Bentham (1748-1832) defended, as the first Western philosopher, animal rights by saying: "The question is not, Can they reason ?, nor Can they talk? But, Can they suffer? Why should the law refuse its protection to any sensitive being?" (3) Bentham, however, did not defend the killing of animals, but only their suffering.
- On the other hand, Kant, the founder of the deontological school, believed that man has no direct moral obligations to animals but

that cruelty to animals leads to cruelty to people. Kant accepted that animals could suffer, justifying their suffering for research purposes but not for sports. In other words, cruelty to animals could be accepted if the benefit to humans outweighs the negative aspects of such an experiment concerning themselves (humans) (4).

On the other hand, animal welfare is based on what animals themselves feel and is defined through the rule of five freedoms (freedom from hunger, thirst, distress, the discomfort of injuries, and freedom to express their natural behavior) (5, 6).

Russell and Birch defined the 3R rule in their book: "The Principles of Humane Experimental Technique", published in 1959 (7, 8). WMS Russell (1925–2006) and RL Burch (1926–1996) were students of Charles Westley Hume (1886-1981) who founded the University of London Animal Welfare Society (ULAWS) in 1926, which later became the Universities Federation for Animal Welfare (UFAW). Hume wrote the first textbook dealing with the welfare of experimental animals and inspired Russell and Birch to define 3Rs.

The International Council for the Harmonization of Technical Requirements for Pharmaceuticals for Human Use (ICH), established in Brussels in 1990, defines the Common Technical Document as an agreement to assemble all the Quality, Safety, and Efficacy information on drugs in a standard format (CTD ICH). Since 2003, the CTD ICH has become mandatory for new drug applications in the EU and Japan, and the strongly recommended format of choice for NDAs submitted to the FDA, United States. It consists of five modules that best illustrate the drug development process (Fig. 2).

Drug development is a continuum that covers all phases of drug life, from discovery to withdrawal from the market, and even the period after that, when data on delayed adverse effects are collected (it can take decades). Drug development is divided into two phases: preclinical development and clinical development (Fig. 3).

The preclinical phase of drug development is contained in module 4 of the CTD triangle (Non-clinical study reports) and is summarized in module 2 (Non-clinical overview and Non-clinical summary).

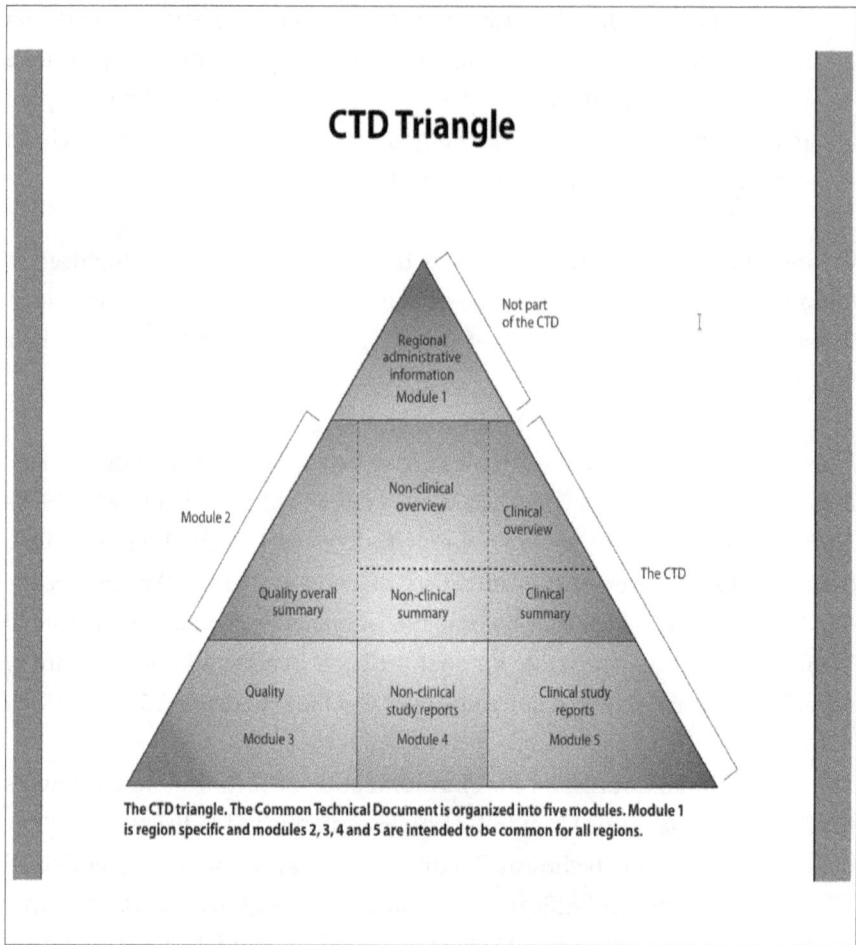

CTD Triangle

The CTD triangle. The Common Technical Document is organized into five modules. Module 1 is region specific and modules 2, 3, 4 and 5 are intended to be common for all regions.

Fig. 2 The CTD triangle (9).

The preclinical drug development phase consists of the following processes: screening, hit identification, and hit-to-lead optimization, and includes essential steps in new substance development, pharmacodynamics testing (including mechanism of action, description of target action, and safety pharmacology), pharmacokinetics (ADME profile) and toxicity (10).

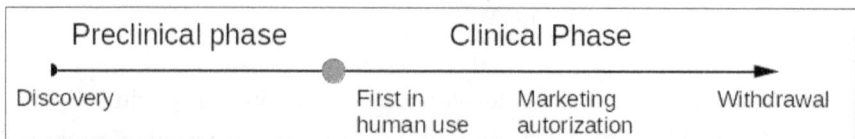

Fig. 3 Drug development process

Defining these phases is essential for understanding all crucial issues for ethical assessment. As we have seen, preclinical drug testing occurs at all levels shown in Figs. 1. However, *in vivo* experiments have a dominant place, not only as the most complex but also as the most serious test of a new substance. The ethical issue at this stage relates to the following:

- obtaining ethical consent for conducting preclinical testing of the new entity;
- selection of the optimal animal model for testing the main pharmacological action, safety pharmacology (indicates the safety profile of the future drug), and pharmacokinetics (selection of the optimal animal species for the most reliable extrapolation of these results to humans);
- toxicological studies, as the most invasive experiments in which the welfare of experimental animals is most disturbed.

European Medicines Agency (EMA) defines Ethical use of animals in medicine testing on its website through the definition of 3Rs, the role of EMA in drug development animal testing, review of the non-EU issues in this field, scientific guidelines, and recommendations on 3Rs in the European Pharmacopoeia (11). In particular, document EMA/470807/2011 of 23 September 2011 entitled: "Statement of the EMA position on the application of the 3Rs (replacement, reduction, and refinement) in the regulatory testing of human and veterinary medicinal products" emphasizes the importance of respect for these principles, as well as all actions organized by the EMA in their implementation (12, 13). Among the EMA scientific guidelines in the field of ethical use of animals in drug testing, the following are specifically mentioned:

- Overview of the current regulatory testing requirements for medicinal products for human use and opportunities for implementation of the 3Rs;
- Overview of the current regulatory testing requirements for veterinary medicinal products and opportunities for implementation of the 3Rs;
- Guideline on regulatory acceptance of 3R (replacement, reduction, refinement) testing approaches;
- Guidance for individual laboratories for transfer of quality control methods validated in collaborative trials with a view to implementing 3Rs.

All guidelines emphasize the importance of respecting the key EU document in the protection of animal welfare, 2010/63/EU (14).

In the USA, the key FDA document regulating the use of animals in testing new drugs is: "Product Development Under the Animal Rule: Guidance for Industry" (15). It emphasizes that "it is unethical to deliberately expose healthy human volunteers to a lethal or permanently disabling toxic biological, chemical, radiological, or nuclear substance ...". However, for animal bioethics, it is important to consider in this document that the FDA accepts testing of new substances on animals only if the following criteria are met:

- There is a reasonably well-understood pathophysiological mechanism of the toxicity of the substance and its prevention or substantial reduction by the product;
- The effect is demonstrated in more than one animal species expected to react with a response predictive for humans, unless the effect is demonstrated in a single animal species that represents a sufficiently well-characterized animal model for predicting the response in humans;
- The animal study endpoint is clearly related to the desired benefit in humans, generally the enhancement of survival or prevention of major morbidity; and
- The data or information on the kinetics and pharmacodynamics of the product or other relevant data or information, in animals and humans, allows selection of an effective dose in humans.

Seemingly, these criteria were introduced primarily to provide reliable information that can be extrapolated to humans. However, they are virtually compatible with 3Rs. Besides, there are obstacles in developing drugs for diseases for which we do not have a reliable experimental model.

These documents are detailed guidelines on conducting preclinical testing of new drugs in *in vivo* trials following all ethical standards. This area of applied ethics developed during the twentieth century as the links between scientific research and industry grew stronger. A counterpart or reflection in the mirror of ethical standards in drug development is ethics in scientific research, which has undergone substantial changes, losing the "innocence" of pure science that is "ethically neutral", and moving from the individual to the collective position. A real boost of ethics in drug development and scientific research occurred during the 1960s, with the promotion of the 3Rs and the establishment of the first ethics committees in human and animal research. Somewhat later, Van Rensselaer Potter (1911-2001) and Peter Singer (1946) eventually reshaped this issue with their publications in which they strengthened global bioethics and continued the work of Fritz Jahr (1895-1953) and Albert Schweitzer (1875-1965). Such an approach in bioethics is holistic and applies to humans and all forms of life and the environment in a continuum and unbreakable chain. In the milieu thus created, Singer could write about the liberation of animals and ask himself (is it just a rhetorical question?) - why do we think we are above animals? Since then, it has been a small step towards giving apes as subjects in the legislation. The practical aspect of the view that apes should be given a special place on the ethical scale is based on the same argument we use when we want to justify using these sublime animals in some experiments: "They are so similar to us!". During the last two decades, a significant shift has been made concerning the use of apes to test some drugs affetcing cognition or to investigate biopharmaceutics (see below). An essential feature of modern bioethics, created in the contact of humanities and biomedical sciences in the 1960s, is the dominance of the utilitarian principle of measuring the benefits and quality of research on the one hand, and risks and harms on the other, which is incorporated into the codes of conduct of all ethics bodies.

Ethical *pros* and *cons* regarding the use of animals in animal experimentation were considered in detail by the Nuffield Council on Bioethics in 2005, without concluding on this controversial issue (16, 17). An independent body in the UK, consisting of scientists, statisticians, philosophers, ethicists, and other stakeholders, chaired by a British baroness, finally agreed that it could not agree or formulate a single conclusion on the issue. Instead, four opposing views were offered, from the standpoint that animal testing is entirely unjustified to the position that everything is allowed without restrictions. The majority of Nuffield Council members would eventually allow animal experimentation if the benefit (for humans) outweighed the damage (to animals), with some absolute limitations (drug testing on great apes). How far this is from the views set out in the Nuremberg Code and the Declaration of Helsinki that medicines must first be tested on animals before testing on humans! Let us mention the topics that the Working Group of the Nuffield Council on Bioethics dealt with in the debate on animal experiments:

• Assessing Pain, Distress, and Suffering in Animals;
• Does Animal Research Lead to Valid Results?
• Is It Morally Acceptable to Cause Pain and Suffering to Animals?
• Can We Ever Agree on Research Involving Animals?, etc.

Choosing a suitable animal model is an essential issue in testing new drugs. Ideally, the animal model should correspond to human disease in terms of:

• similar pathophysiological phenotype (face validity);
• similar etiology (construct validity);
• a similar response to therapy (predict validity).

Often, we do not have such an appropriate experimental model. Sometimes the pathogenesis of the disorder is unclear, the cause is unknown, or the new drug may act by some hitherto undescribed mechanism. As a result, it is difficult to assess the potential benefits of the experiment for humans, and the harm to animals can be significant. Even extrapolating the results obtained from animals to humans does not always have to be straightforward. Good results in the animal model need not get a clinical correlate and vice versa, which should be considered in assessing the ethics of such experiments.

Particular problems are inbred strains obtained after > 20 consecutive sibling matings to design a specific genotype and phenotype (e.g., spontaneously hypertensive rats). There are also transgenic strains with targeted genotype modification. Ethical issues in genetically modified animals are numerous (18). For example, the invasiveness of the experiment can be significant, starting with the surgery necessary to achieve genetic modification (e.g., vasectomy, surgical embryo transfer, and pregnancy-related problems). Also, genetic manipulations require many animals, which collides with the 3R principle of number reduction. Finally, there is the damage itself that occurs due to genetic manipulation. Therefore, scientists no longer only wonder *how* to perform a specific genetic intervention but whether it is *acceptable*.

The ethics of testing drugs that affect cognition and biological drugs in experimental animals require a particular comment. Preclinical pharmacokinetic and toxicological studies may also be an ethical issue.

Cognition

The question arises regarding drugs that affect cognition - to what extent they can be translated or extrapolated to humans. *Cognition* may be defined as a set of psychological processes and activities whose function is knowledge, as opposed to the domain of affectivity (19). An approach to modern cognitive theory, called information-processing, considers the collection, processing, storage, and use of information as elements of the cognition process, comparing the human brain to a computer. The effect of drugs on the cognitive abilities of experimental animals is the subject of the study of neuroethics. Twenty years ago, specific disciplines derived from the unique corpus of bioethics that required a more profound knowledge of biomedical issues, such as gene-ethics, neuro-ethics, and nano-ethics. In mid-May 2002, 150 neuroscientists, bioethicists, doctors of psychiatry and psychology, philosophers, and law and public policy professors gathered in San Francisco and established neuroethics as a discipline (20). Chiaki Kagawa, a Japanese philosopher, called this phenomenon "Balkanization", implying the unnecessary fragmentation of the issue (21). Neuroethics opened up a new, unexplored field of animal bioethics focused on brain functioning.

Experiments with cognition testing were initially performed on primates. The reason was apparent: primates, like humans, receive most of the information through visual signals. Experiments on primates provoke public resistance and are also unacceptable by many research professionals. Neuroscience researchers' arguments that brain investigation is not possible without primates can be rejected because we may use animals more sensitive to acoustic or olfactory signals instead. Namely, rodents (rats and mice) get most of their information through these signals. In the last decades, rodents replaced primates in experiments with cognition. [There is a similar trend in the preclinical development of biopharmaceuticals. Although those drugs are specific to humans and may cause hypersensitivity reactions in animals, testing them on primates is unnecessary. A rough picture of immunogenicity can also be obtained in rodent preclinical studies, although biopharmaceuticals are primarily tested in clinical trials.]

It took a long time to realize that people were not just "big-brained apes", as Charles Darwin claimed (it was thought so until the early 1980s!) (22). Differences in the structure of the human and primate brains are qualitative, not only quantitative, and could be explained in more extensive synaptogenesis. [Either the size of the brain or the existence of specific neurons related to cognition (so-called *Von Economo* or spindle neurons) is not a crucial issue. Such neurons are also present in the brains of great apes, whales, dolphins, and elephants, sometimes more than in humans. These neurons are associated with complex social emotions/cognition, such as empathy, guilt, and shame.] The process of synaptogenesis in the human brain is more complex than in primates. In the former, the peak is reached in the 5th year (prolonged synaptogenesis), and in the chimpanzee only a few months after birth) (23). Specific cognitive abilities of the human brain have their *advantages* (complex social knowledge, language abilities) and *disadvantages* (such as neurodegenerative diseases unique to humans: schizophrenia, autism, and Alzheimer's disease).

Solving several essential dilemmas is vital when dealing with cognition experiments, and above all - whether there is a simple continuity between animals and humans in terms of cognitive abilities and the cognitive abilities of animals. Some empirical facts speak in favor of continuity: all vertebrates (including humans) are sentient and conscious beings, capable of feeling and remembering unpleasant and pleasant experiences; animals may show different abilities similar to humans; animals can predict the order, outcome, and consequences of events, and share with humans similar or the same forms of behavior and strategy, and so on. (24).

Human and animal cognitive processes can be analyzed individually, for example, teaching, short-term memory, causal reasoning, planning, deception, transitive interference, and language. The differences between humans and animals in most of these issues are qualitative, not just quantitative, so there is no simple continuity. However, from a bioethical point of view, such models are justified and can be accepted for neuro- and psychopharmacological experiments.

Preclinical pharmacokinetics

The ethical principle that drugs should be tested on lower species (see previous text) cannot always be applied in preclinical pharmacokinetics. The results obtained on experimental animals are allometrically scaled to extrapolate to humans (recalculation of values according to the ratio of body weight of the tested species). Such a method applies to parameters that are closer to anatomical size or physiological processes (volume of distribution and clearance, respectively) and those that have time dimensions (half-life of drugs). It is necessary to find a suitable pharmacokinetic model to extrapolate data obtained from experimental animals to humans; sometimes, rats are the most appropriate, and for another parameter, the dogs or apes are better.

Preclinical toxicological studies sometimes expose experimental animals to great suffering. The improved statistical methodology has reduced the number of animals that need to be sacrificed from over a hundred to less than ten to calculate the mean lethal dose (LD50). However, the most significant contribution to animal bioethics in this area has been introducing the principles of Good Laboratory Practice (GLP), mandatory for toxicological studies. These principles were established by the Organization for Economic Co-operation and Development (OECD) for allowing the mutual recognition of data. In other words, compliance with GLP standards allows for the verifiability of other people's results and eliminates the need for unnecessary duplication of toxicological studies (25).

Finally, significant advances in testing new drugs regarding 3Rs are refining alternative models. The European Union Reference Laboratory for Alternatives to Animal Testing - EURL ECVAM was established in the early 1990s in Italy and gained prominence when the REACH initiative (the Registration, Evaluation, Authorization and Restriction of Chemicals, REACH; EC 1907/2006) was adopted, according to which chemical substances had to be toxicologically tested in order to increase safety for humans (26, 27). If classic *in vivo* toxicology tests were performed, the cost of testing would be more than a billion euros. The discovery and validation of alternative models can eliminate in vivo toxicological experiments. For example, the Draize test (an acute ocular toxicity test established in 1944) has today been replaced by *in vitro* cell culture testing, which is ethically justified and more reliable and accurate than the Draize test.

In conclusion, bioethical principles are incorporated into all components of preclinical drug development and thus into modern pharmacology. The reasons are numerous, and the benefits are multiple, not only for experimental animals but also for society.

References

1. Bickel MH. Die Entwicklung zur experimentellen Pharmakologie 1790-1850 [The development of experimental pharmacology 1790-1850]. Gesnerus Suppl 2000; 46: 7-158. German.
2. Fox JG, et al. (eds). Laboratory Animal Medicine (3rd ed). Academic Press (Elsevier Inc.): 2015.
3. Bentham J. An Introduction to the Principles of Morals and Legislation, Vol. 1 of 2 (Classic Reprint). Forgotten Books, 2017.
4. Korsgaard CM. Medical Research on Animals and the Question of Moral Standing. Center for Bioethics, Harvard Medical School, 2020. Available at: https://bioethics.hms.harvard.edu/journal/animal-moral-standing, October 2021.
5. Webster J. Animal Welfare: Freedoms, Dominions and "A Life Worth Living". Animals (Basel) 2016; 6(6): 35. doi: 10.3390/ani6060035.
6. Todorović Z. Prostran M, Turza K (eds). Bioethics and Pharmacology: Ethics in Preclinical and Clinical Drug Development. Kerala, India: Transworld Research Network, 2012.
7. Russell WMS, Burch RL. 1959. (as reprinted 1992). The principles of humane experimental technique. Wheathampstead (UK): Universities Federation for Animal Welfare.
8. Tannenbaum J, Bennett BT. Russell and Burch's 3Rs then and now: the need for clarity in definition and purpose. J Am Assoc Lab Anim Sci 2015; 54(2): 120-32.
9. The CTD Triangle. Available on: https://admin.ich.org/sites/default/files/2021-02/CTD_triangle_color_Proofread.pdf.
10. Samuele A, Crespan E, Garbelli A, Bavagnoli L, Maga G. The power of enzyme kinetics in the drug development process. Curr Pharm Biotechnol 2013; 14(5): 551-60. doi: 10.2174/138920101405131111105023.
11. European Medicines Agency (EMA). Ethical use of animals in medicine testing. Available at: https://www.ema.europa.eu/en/human-regulatory/research-development/ethical-use-animals-medicine-testing.

12. Statement of the EMA position on the application of the 3Rs (replacement, reduction and refinement) in the regulatory testing of human and veterinary medicinal products (EMA/470807/2011 as of 23 September 2011). European Medicines Agency, 2011. Available at: https://www.ema.europa.eu/en/documents/other/statement-european-medicines-agency-position-application-3rs-replacement-reduction-refinement_en.pdf.

13. Joint CVMP/CHMP Working group on the Application of the 3Rs in Regulatory Testing of Medical Products (Biennial report 2016/2017). European Medicines Agency, 2017. Available at: https://www.ema.europa.eu/en/documents/report/biennial-report-joint-cvmp/chmp-working-group-application-3rs-regulatory-testing-medical-products-2016/2017_en.pdf.

14. DIRECTIVE 2010/63/EU OF THE EUROPEAN PARLIAMENT AND OF THE COUNCIL of 22 September 2010 on the protection of animals used for scientific purposes. Available at: https://eur-lex.europa.eu/legal-content/EN/TXT/PDF/?uri=CELEX:32010L0063&from=EN.

15. Product Development Under the Animal Rule Guidance for Industry. U.S. Department of Health and Human Services, Food and Drug Administration, Center for Drug Evaluation and Research (CDER), Center for Biologics Evaluation and Research (CBER), October 2015. Available at: https://www.fda.gov/media/88625/download.

16. The ethics of research involving animals. Nuffield Council on Bioethics, 2005. Available at: https://www.nuffieldbioethics.org/wp-content/uploads/The-ethics-of-research-involving-animals-full-report.pdf.

17. Perry P. The ethics of animal research: a UK perspective. ILAR J 2007; 48(1): 42-6. doi: 10.1093/ilar.48.1.42.

18. Ormandy EH, Dale J, Griffin G. Genetic engineering of animals: ethical issues, including welfare concerns. Can Vet J 2011; 52(5): 544-50.

19. Cognition. Larousse dictionaire. Available at: https://www.larousse.fr/dictionnaires/francais/cognition/17005

20. Neuroethics: Mapping the Field. Dana Foundation. Available at: https://dana.org/article/neuroethics-mapping-the-field/

21. Kagawa C. [Neuroethics and bioethics--implications of Balkanization controversy]. Brain Nerve 2009; 61(1): 11-7. Japanese.

22. Darwin C. Descent of Man. J Murray: London, 1871.

23. Somel M, Liu X, Khaitovich P. Human brain evolution: transcripts, metabolites and their regulators. Nat Rev Neurosci 2013; 14(2): 112-27. doi: 10.1038/nrn3372.

24. Premack D. Human and animal cognition: continuity and discontinuity. Proc Natl Acad Sci U S A 2007; 104(35): 13861-7. doi: 10.1073/pnas.0706147104.

25. OECD Series on Principles of Good Laboratory Practice (GLP) and Compliance Monitoring. Available at: https://www.oecd.org/chemicalsafety/testing/

ETHICAL ASPECTS OF TRANSLATIONAL RESEARCH IN BIOMEDICINE

Dragan Hrnčić

Belgrade University, Faculty of Medicine, Institute of Medical Physiology "Richard Burian"

Abstract

Translational research approach in biomedicine passed a way from nice idea to widely accepted scientific highroad. Major funding bodies in the world, like USA National Institute of Health (NIH) recognized the value of translational biomedical research by dedicating a significant fraction of funding to this kind of research activities. Biomedical research area recognizes different scientific approaches including basic studies, clinical studies and population-based studies, while ultimate goal of translational research is to promote the integration of basic research, patient-oriented research, and population-based research, with its major outcome of improving the health. Translational research transforms discoveries of these differently oriented studies into novel clinical armamentaria. Translational research engages society, implies close collaboration with industry, governmental stakeholders and policy makers. Frequently, it engages a number of scientific disciplines outside of biomedical filed. Thus, translational research in biomedicine is a process which faces a number of ethical challenges and requires close ethical monitoring with a lot of specifics. Herein, we will discuss how basic ethical principles should be implemented in organization, conducting and applying translational research with special emphasis on ethical issues and challenges that could arise during each step. It will help researchers and other stakeholders in the translational biomedical research area to recognize and prevent unethical conduct of this kind of research.

Keywords: basic science, clinical medicine, translational imperative, ethics, management

Introduction

From the time of Hippocrates until today, medicine, as a scientific discipline and profession, has always had the same goal, which is to improve and preserve human health. Rising to the level of science, medicine over time has resorted to the application of scientific methods in gaining knowledge about human life, the functioning of the organism, the mechanisms of diseases, as well as ways to cure, alleviate and prevent diseases. In words of the profession, scientific research was needed to understand physiological processes, pathophysiological mechanisms of disease, finding pharmacological therapeutic approaches - development of drugs and medical devices, improving surgery and adequate ways to prevent disease at the individual and population levels.

The challenges posed to medicine became more complex over time, knowledge about health and disease increased, and thus intricate the research process in the field of medicine, the burden of which exceeded the level of individual researcher, scientific genius, innovator and inventor. As in manufactures, the unique mosaic of the human body and its processes is broken down into topography-limited disciplines, as well as different approaches from subcellular and cellular, over the levels of organ, individual organism to population. A large number of different disciplines has become inevitable, and such an approach has brought significant progress in generating scientific knowledge, we can now say, in the field of biomedical sciences.

Biomedical research area recognizes today different scientific approaches including basic studies, clinical studies and population-based studies. In the view of clinical perspective, the imperative is to diagnose and cure already diseased patient. On the other hand, population – based studies strive to gain knowledge on interventions influencing the whole population. From the perspective of basic studies, focus is on understanding mechanisms and finding targets for developing or refining therapeutic armamentaria, or biomarkers that will facilitate early diagnosis[1]. Interdisciplinary approach is broadly defined as "a mode of research by teams or individuals that integrates information, data, techniques, tools, perspectives, concepts, and/or theories from two or more disciplines or bodies of specialized knowledge to advance fundamental understanding or to solve problems whose solutions are beyond the scope of a single discipline or area of research practice"[2]. The concept of interdisciplinary and multidisciplinary presumes horizontal integration and cooperation. Translational approach in biomedical research struggles to integrate scientific approaches vertically.

Ultimate goals of translational research are to promote the integration of basic research, patient-oriented research, and population-based research, with its major outcome of improving the health. Translational research transforms discoveries of these differently oriented studies into novel clinical armamentaria. Translational research engages society, implies close collaboration with industry, governmental stakeholders and policy makers. Frequently, it engages a number of scientific disciplines outside of biomedical filed. Thus, translational research in biomedicine is a process which faces a number of ethical challenges and requires close ethical monitoring with a lot of specifies. In this chapter, we will discuss how basic ethical principles should be implemented in organization, conducting and applying translational research with special emphasis on ethical issues and challenges that could arise during each step.

Translational research roots and incentives

Although the idea and concepts of translational research are recognized much earlier, the wording translational research was introduced in early 1990s. However, the major impetus for translational research was the realize of NIH "Roadmap for Medical Research" in 2003.[3] Mission of NIH is "to seek fundamental knowledge about the nature and behavior of living systems and the application of that knowledge to enhance health, lengthen life, and reduce illness and disability". The vast majority of funding resources has been allocated to translational research, and other funding bodies, public and private, also followed the idea of translational research. A number of scientific institutions and scientists responded to the calls and reshaped their practices to respond to the initiative that has so many appreciated values and proclaimed targets that should be achieved for the benefits of the individual patients, healthy people and society that actually funds research activities. Number of publications in this framework increased. Further historical considerations could be found elsewhere.[4]

Defining translational research

Before we cite some of the most common definitions of translational research in biomedical sciences, we should be aware of the fact that these definitions evolved over time in order to include all the steps of translational approach to vertical integration.

Translational research could be defined in several ways. Wang[5] defined it as "research (which) transforms scientific discoveries arising from laboratory, clinical or population studies into new clinical tools and applications that improve human health by reducing disease incidence, morbidity and mortality"[5]. From the perspective of translational physiology (the American Physiological Society), translational research is "the transfer of knowledge gained from basic research to new and improved methods of preventing, diagnosing, or treating disease, as well as the transfer of clinical insights into hypotheses that can be tested and validated in the basic research laboratory".[6] According to European Society for Translational Medicine, translational medicine is "interdisciplinary branch of the biomedical field supported by three main pillars: benchside, bedside and community. The goal is to combine disciplines, resources, expertise, and techniques within these pillars to promote enhancements in prevention, diagnosis, and therapies"[7].

Rubio *et al.*[8] stressed the importance of multidirectional integration and introduced the health of the public as the final outcome of translational research. In their words: "Translational research fosters the multidirectional integration of basic research, patient-oriented research, and population-based research, with the long-term aim of improving the health of the public"[8]. This definition is particularly applicable for further discussions on ethical aspects of translational research.

Be aware that, still, there is no single widely accepted definition of the translational research and that a number of models exist[9]. Herein, we will elaborate these models very briefly.

Critical steps in translational research: phases and gaps

The earliest model of translational research has been proposed by Sung et al.[10] as a two-step model with T1 and T2 phases. T1 phase "involves the transfer of new understandings of disease mechanisms gained in the laboratory into the development of new methods for diagnosis, therapy, and prevention and their first testing in humans", while T2 phase "block affects the translation of results from clinical studies into everyday clinical practice and health decision making"[10]. Later, models of Westfall et al.[11], Dougherty and Conway[12] with T1, T2, T3 have been proposed. Khoury et al.[13] proposed model with T1 to T4.

In model proposed by Callard, Rose and Wykes[14], which also is four – phase model, one of the key elements is inclusion of public in planning the translational research.

Hence, from initial two main phases of T1 being bench-to-bedside (proof of concept, clinical trials phase I and II) translation from laboratory findings to clinical research and T2 being oriented toward translation into clinical practice (improving health, encompassing phase III clinical trials etc.), we are including now T4- related to translation to public health, public health policies, prevention, behavioral and lifestyle changes.

Nowadays T5 has been frequently added to, we could freely say, translational continuum. Namely, T5 strives to translate the maximum benefits of scientific and medical innovation to global populations worldwide and extends toward social health policies[15]. Therefore, Waldman and Terzic elaborated that "T5 translational teams must incorporate investigators with knowledge domains that extend beyond the laboratory and clinic, engaging political and social scientists, engineers, economists, anthropologists, and population biologists to define the critical path that maximizes investments in research for the good of the global".[16] Herein, we would like to add the bioethicist as an inevitable member of the team, having in mind all the ethical issues closely related to translational research.

It should be noted that bidirectional link exists between the phases. In the words of physiological vocabulary, translational continuum should be viewed as a chain of feedback loops rather than a linear process (Fig. 1). This is of high importance with the aim to secure better communication between those involved in different phases of translational research and to finely tune each process. Some examples of translational research view on physiology topics could be find elsewhere.[17]

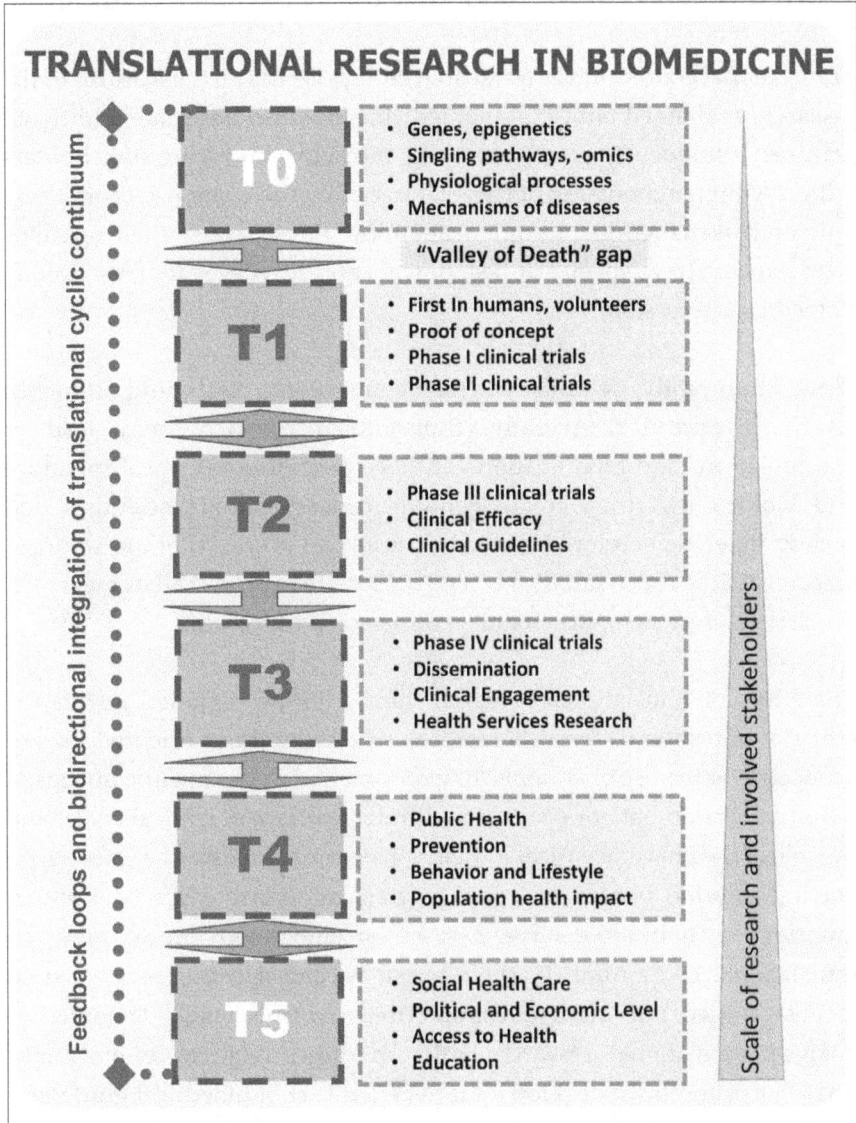

TRANSLATIONAL RESEARCH IN BIOMEDICINE

- T0
  - Genes, epigenetics
  - Singling pathways, -omics
  - Physiological processes
  - Mechanisms of diseases

"Valley of Death" gap

- T1
  - First In humans, volunteers
  - Proof of concept
  - Phase I clinical trials
  - Phase II clinical trials

- T2
  - Phase III clinical trials
  - Clinical Efficacy
  - Clinical Guidelines

- T3
  - Phase IV clinical trials
  - Dissemination
  - Clinical Engagement
  - Health Services Research

- T4
  - Public Health
  - Prevention
  - Behavior and Lifestyle
  - Population health impact

- T5
  - Social Health Care
  - Political and Economic Level
  - Access to Health
  - Education

Feedback loops and bidirectional integration of translational cyclic continuum

Scale of research and involved stakeholders

Figure 1. Translational research in biomedicine as a chain of feedback loops. Classical and additional phases with gaps. Based on models described elsewhere.[9, 16, 18, 22]

Specific gaps also exist between specific phases. The most significant gap is between T1 and T2 and is widely known as "volley of death" since the majority of developments made in the laboratories (i.e., drug discoveries) do not continue their life in T2.[18]

Discrete or integrated ethical view on translational continuum

Translational continuum as a research enterprise has all the features of the research, as defined entity, in biomedicine, but also has some additional features. It frequently goes beyond the medicine, involving stakeholders within within and outside of the health sector. Therefore, we cannot rely only on research ethics and medical ethics. Physicians in their relations with patients have special duties, powers and privileges that are unique to medical profession.[19]

More importantly, it addresses some questions that could influence all living creatures. Actually, translational research "may lead to manipulations and modifications of those processes so critical to nature and biology that they create genuine concern among scientists and society alike."[20] Therefore, we need close ethical oversight of translational research that will also satisfy the needs of society to protect all its members: researchers, research participants, end-users of outcomes.

The ethics of translational research should not be regraded just as the ethics of translational medical research which should be oriented toward "concerns on the use of animals for experimental or therapeutic purposes, requirements on patient consent at different phases of clinical trials, and the role of the pharmaceutical industry in directing innovative research."[21] Having in mind phases and gaps in translational research continuum, question is should we have discrete or integrated ethical view on translational continuum. In their recent review, Hostiuc et al.[22] tried to answer two important questions, the first one being "Should we analyse translational research ethics in direct relation to the steps/phases, or should we develop a more general set of bioethical guidelines guiding the entire translational research process?" and the second being "Should we consider translational research as a subtype of biomedical research, or should we develop a new set of bioethical tools specifically directed to the often transdisciplinary characteristics of translational research?" According to them we should emphasize the unique features of translational research which makes it different in comparison with other biomedical research when framing the ethics applicable in this filed. Comprehensive ethical analysis of translational research should be focused on entire translational continuum with special attention on translational gaps.

We do agree with Hostiuc et al.[22] that ethics of translational research could not be viewed only through the prism of research ethics or only through the prism of medical ethics, but should going much beyond taking into account translational phases, data and knowledge transfer and the unique features of translational research continuum itself.

Basic and critical ethical principles relevant to translational research

Translational research should be also assessed through the prism of basic ethical principles[23] which are bound in a number of ethical codes related to medical practice and research. These *prima facie* principles are: i) respect for autonomy, ii) nonmaleficence (*primum non nocere*), iii) beneficence, and iv) justice. It should be stressed out that these principles have equal values and only balanced and justified form "an analytical framework of general norms derived from the common morality that form a suitable starting point for biomedical ethics".[23] Further elaboration of these principles could be find in relevant ethical literature. Principles of respect for persons, beneficence, and justice are incorporated in the Belmont Report, the National Commission on Protection of Human Subjects in Biomedical and Behavioral Research. [24] Koski[20] argued that these principles "provide the normative basis for the responsible scientist engaged in human subjects' research, and any scientist unwilling or unable to be guided by them should not be permitted by society or his peers to participate in human research". Hence, scientist that will be engaged in translational research have to be trained in basic bioethics and to constantly not just know the principles, codes, directives, but most importantly to internalize them. The most significant principles to analyze, from the perspective of translational research are beneficence and justice, but not neglecting the risk analysis.

Namely, all the stakeholders included in the translational research should strive to achieve balance between the risks and benefits; risks for the participant and society; benefits for participants, population and society. It is important to be aware that risk benefit ratio analysis should not be performed solely for particular phases of translational research, but rather for translational research integrally. Moreover, just distribution of gained benefits elaborated in more details by Goering et al.[25] should be secured. As it has been nicely formulated[26] "central issues when designing translational efforts should be whether positive steps are being taken to prevent harm, promote good, and promote the just distribution of benefits and burdens of research."

In-depth analysis of the risks related to the process of translational research, as well as to its outcomes (products, services, technologies, policies) and consequences should be responsible done in advance using a consequentialist approach. Risk management should be related to prevention and mitigation of the risks for the subjects, but not only to them, but also to society and population as a whole.

Basic ethical principles and norms that are required and suggested to be internalized by all the biomedical scientist, should be also present in translational research and scientist conducting translational research should be aware of all specificities that characterized translational research. Among these, societal concerns have to be well addressed and trust in science and medicine has to be preserved. Koski[20] suggested following six principles relevant to guide translational scientists:

1. "The questions asked and the approaches proposed to answer them should be soundly justifiable both ethically and scientifically to one's scientific peers as well as to a reasonable, well-informed public"

   Indeed, ethically sound research must be also scientifically sound. It must be approved by scientific community, but also by society, i.e., member of society that is well informed. Informed consent is highly valuable ethical legacy, but its extension to informed society is necessary in translational endeavor.

2. "No scientific studies that may seriously harm the environment, individuals, groups of individuals, or populations at large should be undertaken unless the risks are predictable and controllable."

Prima facie principle of nonmaleficence, in medical profession known also as *primum non nocere* or do not harm should not be related only to the participants, research subjects. Risks has to be assessed also against environment and population.

3. "Particular caution must be exercised when consequences of the proposed research are not reversible."

   In this situation, balancing risks and benefits is of particular importance. Such decisions require broad consensus.

4. "When proposed experiments will use living creatures as a means to advance science, the use of animal or human subjects should be justified and the subjects should be treated with dignity and respect."

   Prima facie principle respect for autonomy with focus on paternalistic approach to research subjects should always be considered in designing and realization of translational research. Informed consent as an institution and a process must be respected rigorously, especially in translation research using genetic material.[27] On the other hand, it has been recognized that traditional means for informed consent has to be readdressed in the environment of translation research, especially in large scale studies when health literacy of potential participants are questionable. Therefore, concepts of so-called dynamic consent using IT tools has been proposed[28] for application in translational research. Animal welfare has to be secured and all the principles in working with animal subjects have to be followed up.

5. "Knowledge acquired through scientific studies should be shared and used in a manner that optimizes its utility and benefit for all."

   Principles of responsible conduct of research are applied to translational research without exception. This is particular important for knowledge, information and data transfer that is a process which characterizes translational research, i.e., transitions in the chain of translation. It should be clear that data transfer is not the same as data sharing, since the first is related to internal flow of data and information, while the later one is related to spreading the results among scientific community and beyond. Data management is recognized as ethically and technically challenged and some digital solutions has been proposed.[29] Research misconduct (falsification, fabrication, etc.) have to be prevented and data confidentiality strictly monitored. Trustworthiness should be attained and preserved. Research integrity has to be the primary duty at individual and institutional level in creating an environment that promotes responsible conduct of research.[30]

Data sharing should be according FAIR (Findable, Accessible, Interoperable, and Reusable) principles, but with keeping in mind risk-benefits of its possible misuse.

6. "Scientists must be willing to be personally accountable for the quality and integrity of their work and its consequences."

Therefore, scientists have to be independent in their decisions and to act in such a way to perform the best scientific practice. As in authorship criteria, accountability, not just responsibility, has to be overtaken by scientist.

Specific issues to be addressed in translational research

Translational research is also connected with some issues among which the most discussed are: sampling bias, population representation and conflicts of interest in collaboration with industry that prioritize the production of products and profit, and the difficulty for marginalized groups to benefit from publically funded research.[26]

Conflict of interest on institutional level is particularly sensitive issue in framework of translational research since the final steps in translational research are closely connected to collaboration with industry. Private sector, private funding entities and companies play an essential role in the acceleration of scientific applicability and interdisciplinary collaboration. Therefore, all the principles and strategies to prevent possible conflict of interest should be applied when planning and designing translational research. Special emphasis should be on the role of funding body in design, realization and publication of research results.

Publication of negative results should not be prohibited.

All research lines in biomedicine, especially if they are funded by public resources, should promote health for all, and translational research has to strive to decrease inequalities in health[26]. Therefore, clear voice for translating translational medicine into global health equity has been articulated.[31]

Translational research requires significant amount of material resources which are often scarce and have to be justly distributed to different research topics. Therefore, ethical issues related to allocation of funds are especially frequent in translational research environment. Decision makers should utilize a variety of methods of health economic analysis and struggle to achieve just distribution of resources. Critics are related to the funding policies accusing them to overestimate the powers of translational research and lamenting that fundamental scientific efforts are neglected. When interviewed, scientists (in the field of addiction neurosciences) pointed out following: i) optimism that their own research has potential for translation ii) value of basic science and 3) concerns on pushing translation too quickly.[32]

Many bioethicists pointed out several shortcomings of translational research from special risks, questionable benefits to increased disparity, but at the same time bioethicists should balance the scaling up oversight of translational research.[33] In creating norms and governance policies, these values have to be envisaged: openness, participation, effectiveness, coherence, and accountability.[34] In governing policies, transparency and individual judgments also should be included.[35] Diversity of representatives in governing committees have to be achieved having in mind that a number of stakeholders are involved in translational research, not just researchers with special emphasis on community-engaged research such as community-based participatory research and community-partnered participatory research.[36]

Translational research also evolves with developments of novel technologies. In the recent past, digital medicine also affected translational research. Developments in the field of artificial intelligence, machine learning and algorithms to assess human behaviors using social media and digital public sources, opened new perspectives in translational research and evoked additional ethical issues.[37, 38] Open questions are

"What types of decisions should be delegated to computer -based systems (artificial intelligence)? Who decides the underlying algorithms and decision patterns?"[39]

Education and training of all physicians and other researchers in the arena of translational research has to be primary duty with special emphasis on clearly defined competencies.[36,40]

Conclusion

Translational research approach in biomedicine passed a way from nice idea to widely accepted scientific highroad. Translational research transforms discoveries of differently oriented studies into novel clinical armamentaria. Translational research engages society, implies close collaboration with industry, governmental stakeholders and policy makers. Frequently, it engages a number of scientific disciplines outside of biomedical filed. Thus, translational research in biomedicine is a process which faces a number of ethical challenges and requires close ethical monitoring with a lot of specifies. Having in mind unique features of translation research we need to adopt integrated ethical view on translational continuum. Also. basic ethical principles should be implemented in organization, conducting and applying translational research with special emphasis on ethical issues and challenges that could arise during each phase and gap. All the efforts on individual, institutional and societal level have to be undertaken in order to recognize and prevent unethical conduct of translational research. Bioethics education and training of all stakeholders in the arena of translational research has to be primary duty.

References

1. Hrnčić D. Translational, multidisciplinary and integrative studies in neurophysiology: real need or contemporary fashion in brain research. Med Podml. 2018;69(3):1-2
2. Committee on Facilitating Interdisciplinary Research, Committee on Science, Engineering, and Public Policy. Facilitating interdisciplinary research. National Academies. Washington: National Academy Press, 2004, p. 2.
3. Butler D. Translational research: crossing the valley of death. Nature. 2008;453(7197):840-2.

4.   Maienschein J, Sunderland M, Ankeny RA, Robert JS. The ethos and ethics of translational research. Am J Bioeth. 2008;8(3):43-51.

5.   Wang X. A new vision of definition, commentary, and understanding in clinical and translational medicine. Clin Translat Med 2012;1:5.

6.   Hall JE. The promise in translational physiology. Am J Physiol Endocrinol Metab. 2002;283:E193-E194

7.   Cohrs R, Martin T, Ghahramani P, Bidaut L, Higgins PJ, Shahzad A. Translational Medicine definition by the European Society for Translational Medicine. New Horizons in Translational Medicine. 2015;2:86-8.

8.   Rubio DM, Schoenbaum EE, Lee LS, Schteingart DE, Marantz PR, Anderson KE, Platt LD, Baez A, Esposito K. Defining translational research: implications for training. Acad Med 2010;85(3):470-5

9.   Trochim W, Kane C, Graham MJ, Pincus HA. Evaluating translational research: a process marker model. Clin Transl Sci. 2011;4(3):153-62.

10. Sung NS, Crowley WFJ, Genel M, Salber P, Sandy L, Sherwood LM, et al. Central challenges facing the national clinical research enterprise. JAMA. 2003; 289(10):1278–1287.

11. Westfall JM, Mold J, Fagnan L. Practice-based research— "Blue Highways" on the NIH roadmap. JAMA. 2007; 297(4): 403–406.

12. Dougherty D, Conway PH. The "3T's" road map to transform US health care. JAMA. 2008; 299(19):2319–232.

13. Khoury MJ, Gwinn M, Yoon PW, Dowling N, Moore CA, Bradley L. The continuum of translation research in genomic medicine: how can we accelerate the appropriate integration of human genome discoveries into health care and disease prevention? Genet Med. 2007; 9(10): 665–674.

14. Callard F, Rose D, Wykes T. Close to the bench as well as at the bedside: involving service users in all phases of translational research. Health Expect. 2012;15(4):389-400.

15. Brook RH. Medical leadership in an increasingly complex world. JAMA. 2010; 304: 465–466

16. Waldman SA, Terzic A. Clinical and translational science: from bench-bedside to global village. Clin Transl Sci. 2010;3(5):254-7.

17. Hrnčić D. Sleep, nutrition and physical exercise in brain hyperexcitability: translational viewpoint. Belgrade: Andrejevic Ed. 2015

18. Seyhan, AA. Lost in translation: the valley of death across preclinical and clinical divide – identification of problems and overcoming obstacles. Transl Med Commun 2019: 4- 18

19. Rhodes, R. The Trusted Doctor: Medical Ethics and Professionalism. Oxford, UK: Oxford University Press, 2020.

20. Koski G. Chapter 24 - Ethical Issues in Translational Research and Clinical Investigation, Eds: David Robertson, Gordon H. Williams, Clinical and Translational Science (Second Edition), Academic Press, 2017

21. Bærøe K. Translational ethics: an analytical framework of translational movements between theory and practice and a sketch of a comprehensive approach. BMC Med Ethics. 2014;15:71.

22. Hostiuc S, Moldoveanu A, Dascălu MI, Unnthorsson R, Jóhannesson ÓI, Marcus I. Translational research-the need of a new bioethics approach. J Transl Med. 2016;14:16.

23. Beauchamp T, Childress J. Principles of biomedical ethics (7th ed.). New York: Oxford University Press, 2013.

24. National Commission for the Protection of Human Subjects of Biomedical and Behavioral Research. The Belmont report: Ethical Principles and Guidelines for the Protection of Human Subjects of Research. Office of the Secretary, Health and Human Services 1979

25. Goering S, Holland S, Edwards K. Making Good on the Premise of Genetics: Justice in Translational Science. In Achieving Justice in Genomic Translation: Re- Thinking the Pathway to Benefit. Oxford University Press, USA, 2011

26. Dong Y. Translational Research: Ethical Considerations. Sound Decisions: Undergrad Bioethics J 2017:3(1):Article 2.

27. Halverson CME, Bland ST, Leppig KA, Marasa M, Myers M, Rasouly HM, Wynn J, Clayton EW. Ethical conflicts in translational genetic research: lessons learned from the eMERGE-III experience. Genet Med. 2020;22(10):1667-1672

28. Jacquier E, Laurent-Puig P, Badoual C, Burgun A, Mamzer MF. Facing new challenges to informed consent processes in the context of translational research: the case in CARPEM consortium. BMC Med Ethics. 2021;22(1):21.

29. Gu W, Yildirimman R, Van der Stuyft E. Data and knowledge management in translational research: implementation of the eTRIKS platform for the IMI OncoTrack consortium. BMC Bioinformatics 2019: 20:164. https://doi.org/10.1186/s12859-019-2748-y

30. Petrini C, Minghetti L, Brusaferro S. A few ethical issues in translational research for medicinal products discovery and development. Ann Ist Super Sanita. 2020;56(4):487-491.

31. Isaacson Barash C. Translating translational medicine into global health equity: What is needed? Appl Transl Genom. 2016;9:37-9.

32. Ostergren JE, Hammer RR, Dingel MJ, Koenig BA, McCormick JB (2014) Challenges in Translational Research: The Views of Addiction Scientists. PLoS ONE 9(4): e93482.

33. Sofaer N, Eyal N. The diverse ethics of translational research. Am J Bioeth. 2010;10(8):19-30.

34. Lopez de la Vieja MT. Ethics and governance in translational research Ethics Med Pub Health 2016;2(2):256—262.

35. Riva L, Petrini C. A few ethical issues in translational research for gene and cell therapy. J Transl Med. 2019;17(1):395.

36. Khodyakov D, Mikesell L, Schraiber R, Booth M, Bromley E. On using ethical principles of community-engaged research in translational science. Transl Res. 2016;171:52-62.e1.

37. Topol EJ. A decade of digital medicine innovation. Sci Transl Med. 2019;11(498):eaaw7610.

38. Char DS, Shah NH, Magnus D. Implementing Machine Learning in Health Care - Addressing Ethical Challenges. N Engl J Med. 2018;378(11):981-983.

39. Kimberly RP. Translational Research-For the Individual and the Community. J Health Care Poor Underserved. 2019;30(4S):79-85.

40. Arnold JF, Boan AD, Lackland DT, Sade RM. Clinical and Translational Research Ethics: Training Consultants and Biomedical Research Personnel. Am J Bioeth. 2018;18(1):57-61.

EXPERIMENTAL MODELS OF ACUTE AND CHRONIC PAIN

Sonja Vučković, Dragana Srebro, Katarina Savić Vujović

University of Belgrade, School of Medicine, Department of Pharmacology, Clinical Pharmacology and Toxicology

Abstract

In recent years, there is increasing concern that traditional experimental models of pain are not sufficiently predictive in the development of new drugs for the treatment of pain. A number of targets identified in animal studies have not been confirmed in clinical trials and only a small part of preclinical studies have been translated to successful clinical trials. The lack of availability of new pain treatment strategies has been linked to a failure of existing animal models to be similar to clinical pain. All of these support the view that new models utilizing novel endpoints are needed in order to overcome the current translational barrier in the development of novel pain drugs. Currently, several recommendations are offered to overcome the translational barriers in the development of novel pain drugs. The International Association for the Study of Pain (IASP) announced 2022 as the Global Year for Translating Pain Knowledge to Practice. The second barrier in animal pain research is ethical dilemma. The researchers should be aware that animals suffer from pain, like humans and therefore deserve moral respect and legal protection. The researchers should follow the national and regional guidelines and policies concerning ethics in animal research and work under the supervision of local Institutional Ethics Committees.

Keywords: experimental pain models, translation, ethics

Introduction

Pain is one of the most common symptoms of the disease and determining the origin of pain may help in making diagnosis and direct the course of treatment. Despite intensive preclinical and clinical research efforts over the past few decades there have been limited successes in the discovery and development of new analgesics[1]. Furthermore, the management of pain in both acute and especially chronic settings is still far from optimal. Today, all known analgesics have significant adverse effects. Acute pain is a common and transient experience and is usually successfully treated with short courses of non-opioid or opioid analgesics. However, chronic pain (CP) is often difficult to treat. CP is defined as pain that lasts more than 12 weeks and is linked to lower quality of life (sleep disturbance, depression, fatigue), lost productivity, and increased healthcare costs[2]. It is estimated that CP affects more than 30% of people worldwide[3]. Recently, International Association for the Study of Pain (IASP) provided the classification of CP for the International Classification of Diseases (ICD-11) establishing chronic pain as a health condition[4]. Opioid use for non-cancer CP remains controversial given its uncertain effectiveness and harms due to diversion, addiction, overdose, and death[5, 6]. The current crisis of opioid analgesic in United States, Canada, and Australia requires the urgent discovery of new analgesics[7].

Animal pain research contributes to a better understanding of the mechanisms underlying pain generation, identifying new analgesic targets, and the development of drugs in the treatment of pain in human and veterinary medicine. Intact animals have a perception of pain and behave in a similar way as humans who experience pain[8-11]. However, unlike humans, animals cannot verbalize the experience of pain. Therefore, the measurement of pain in animals is indirect, mainly based on behavioral (paw or tail withdrawal reflex, vocalization, biting, paw licking, reduced activity or inactivity, analgesic position)[12-17], and less often neurovegetative reactions to pain (increase in blood pressure, acceleration of pulse rate)[18]. Models of experimental pain in animals are designed to mimic as much as possible different painful conditions in humans and evaluate the usefulness of new analgesics in these different conditions. Animal model data are used to determine efficacious exposure levels and preclinical safety margins[19]. They also provide dose ranges in early human studies and are required prior to initiating human trials[9].

In general, two important issues are involved in the use of animals in pain research. The first one is how to translate results obtained from preclinical animal pain studies into humans[20, 21] and the second is how to conduct responsible and human pain research on animals[22, 23].

The present paper is aimed to discuss: 1) recent advances that should improve the relevance and translatability of animal-to-human findings and 2) ethical issues in animal pain research.

Translation in pain research

In recent years, animal behavioral models, have been increasingly tested and have proven successful in elucidating pathophysiological mechanisms of CP but have failed to deliver clinical advances needed to treat patients with CP[20, 24]. While preclinical efficacy of several drugs (ziconitide, tanezumab, TRPV1 antagonist compound[25-27] has been recently evaluated and successfully translated to the clinic, there are a lot of examples where efficacy demonstrated in animal models of pain has failed to translate to clinical efficacy[28-31]. It has been reported that only 1 of 10 experimental analgesics get approval to pass to Phase I clinical trial[32], and that failure rates in the clinical phase is around 90 to 95%[33]. It should also be acknowledged that the basic strategy of research including traditional preclinical models and clinical trials has not changed significantly in over 20 years[34]. All of these support for the view that new models utilizing novel endpoints are needed in order to overcome the current translational barrier in the development of novel pain drugs20,[35](Table 1). Finally, IASP has named 2022 the Global Year for Translating Pain Knowledge to Practice[36].

Nociception vs. pain

Nociception

Nociception (from Latin nocere 'to harm or hurt') is the neural process of encoding and processing noxious stimuli[37]. Nociception refers to a signal arriving at the central nervous system as a result of the stimulation of specialized sensory receptors in the peripheral nervous system called nociceptors. Nociception is important for the "fight or flight response" of the body and protects us from harm in our surrounding environment. Noxious stimulation triggers nocifensive behavior (withdrawal reflexes), physiological responses (increases in heart rate and blood pressure), as well as unpleasant emotional state of pain. Many of these responses can also occur in organisms that do not experience pain (e.g, anesthetized animals, or those with spinal lesions that prevent nociceptive information from reaching higher central nervous system structures). Some nociceptive responses (e.g, withdrawal reflexes in spinal cord-transected animals) do not necessarily indicate pain[8,9].

Pain

Pain and nociception are different phenomena[37]. Pain is mediated through the activity of nociceptors, and depends on the interaction between those nociceptors and higher processing centers in the brain to generate the negative emotional component associated with the potential harm. Nociception may occur when someone is unconscious, whereas pain by definition cannot[19]. Pain is an experience of the conscious brain, a sensory and emotional percept.

According to IASP, pain in humans is "An unpleasant sensory and emotional experience associated with actual or potential tissue damage, or described in terms of such damage[38]. Zimmermann[39] re-interpreted the IASP definition of pain so that it could be applied to animals: "an aversive sensory experience caused by actual or potential injury that elicits progressive motor and vegetative reactions, results in learned avoidance behavior, and may modify species specific behavior, including social behavior".

Pain in humans is an unpleasant signal to us that something hurts. It is a complex experience and differs greatly from individual to individual, even between those with similar injuries and/or illnesses. Pain in humans is a multidimensional subjective experience comprising sensory, motor, cognitive, autonomic and affective responses, and it is often reported by patients with reference to such responses[40]. However, animals could not verbally articulate their pain. Verbal description is only one of several behaviors to express pain; inability to communicate does not negate the possibility that a human or a nonhuman animal experiences pain. Therefore, an ongoing challenge in laboratory animal research is to determine whether responses that could merely be nociceptive are also indicative of pain, and, conversely, whether the abolition of nociceptive responses indicates the successful abolition of pain[10, 11]. The distinction between nociception and pain is not always clear. Pain is subjective in both animal and humans, and may vary among individuals due to previous painful experience, sex differences, or clinical comorbidities[35]. Therefore, pain cannot be "seen" by anyone other than the person experiencing it. In addition, there are multiple neurobiologically and phenomenologically distinct categories of pain and the pathophysiological mechanism of many pain syndromes remains unknown. For the time being there are limits to the extent that we can "measure" pain, and say with confidence that we fully understand pain[41, 42].

Acute (nociceptive) pain is a result of direct activation of nociceptors (somatic, visceral). However, chronic pain can involve inflammatory, neuropathic, ischemic, and compression mechanisms at multiple sites. Unlike nociceptive pain, chronic pain occurs in the absence of external stimuli (ongoing or spontaneous pain), from normally innocuous stimuli (allodynia) and with enhanced pain due to normally painful stimuli (hyperalgesia).

Current approach in the treatment of pain is that pain should be treated according to the pathophysiological mechanisms responsible for the generation of the particular type of pain. Therefore, animal models of pain are designed to mimic distinct clinical diseases to better evaluate underlying mechanisms and potential treatments. Also, they are used to rank-order compounds in the process of establishing structure–activity relationships (SAR)[14, 19]. Using these models, researchers can control experimentally induced pain (nature, location, intensity, frequency, and duration of stimulus effects) and quantify psychophysical, behavioral, or neurophysiological responses[10, 11].

Animal models of pain

The animal model of pain is comprised of three basic components: the subjects, the assay, and the outcome measure[10]. Each component requires careful consideration in order to optimize potential translational value to the proposed human pain syndrome being modeled.

Subjects

The main subjects used in preclinical pain studies are rodents, especially rats. Despite difficulties in conditioning in mice, the development of transgenic mice, made them very popular in pain research[10]. The extensive use of rodents is based in the similarities in the neuroanatomy and physiology across mammalian species[10]. In order to overcome current translational barriers in the development of novel drugs, many authors agree that translational pain research must incorporate additional determination of cross-species similarities and differences in the pain-related anatomy and physiology, including biopsychosocial processes too[43] (Table 1).

Only a limited number of rodent strains are used in pain research and they are frequently restricted to young age and one sex to elicit reliable and reproducible responses to represent the pain state of interest. As chronic pain states in humans are more frequent to occur in median-age or older and mainly women patients, some author believed that restricting experimental animals to a certain age and gender do not represent the whole population to which the painful condition is modeled[44]. A growing body of evidence shows a huge impact of gender, age and comorbidity in animals on the development of chronic pain and suggests that these characteristics must be taken into account when researching pain in animals[43, 45] (Table 1). On the clinical side, the lack of sanitization of phenotypes in clinical trials has also contributed to the insufficient success of efforts to translate basic research to the clinic[35, 46]. The phenotypic characteristics recommended for use in clinical research are: psychosocial factors, symptom characteristics, sleep patterns, responses to noxious stimulation, endogenous pain-modulatory processes, and response to pharmacologic challenge[47].

Using large animals in pain research is increasingly popular in recent years and there are many good reasons for this[48] (Table 1). Large animals are phylogenetically closer to humans than rodents and well-validated pain models in non-rodent species could overcome gaps in current rodent pain research and increase the probability of the successful development of new analgesics. Also, this could enhance the speed and reduce the costs of research. Previous reports have strongly suggested the use of veterinary conditions as translational pain models suggesting that naturally occurring chronic pain in animals may offer greater predictivity over induced models of chronic pain in rodents. Some authors even suggest incorporation of a II phase veterinary clinical trial between the preclinical research and human clinical trials. It has also been suggested that naturally occurring models may better reflect the complex genetic, environmental, temporal and physiological influences present in humans[49].

Pain assays

Pain assays explore the pain system under controlled settings. They couple a method for inducing a pain, such as mechanical trauma or injection of an algogenic compounds, to a behavioral assessment (the outcome measure)[50]. Assays vary in the topography of nocifensive responses elicited and the noxious stimuli used to induce such behaviors. Some responses that are measured in traditional assays occur in spinalized animals and therefore are thought to be spinal withdrawal reflexes, whereas other induced behaviors involve the recruitment of supraspinal central nervous system (CNS) regions[14]. From a clinical point of view, it is important that an animal model of pain can predict the effectiveness of a potentially new analgesic in humans. Of note, it should be emphasized that pharmacological mechanisms that modulate nociception at the spinal level have shown a very strong correlation to efficacy in humans (opioids, alpha-2 adrenergic agonists, N-type calcium channel blockers, local anesthetics) and that spinal delivery of drugs is an important strategy for management of pain in some cases[51].

According to[52] assays are classified as assays of acute pain, tonic pain, neuropathic pain, assays based on etiology (cancer pain, bone fracture pain, postoperative pain, chemotherapy induced neuropathy, postischemia pain), as well as assays "batteries".

There is a broad range of nociceptive stimuli available to induce acute and chronic pain states. The choice depends on the type of pain you are attempting to model. Each stimulus may be preferred to model a particular pain condition[53]. In general, nociceptive stimuli activate distinct nociceptors in the nervous system and can be grouped into four categories: thermal, electrical, mechanical, and chemical. These four main nociceptive stimuli offer advantages like simple methodology and equipment, easy control of stimulus intensity, little training required, rapidity of measurement, etc. It is important that procedure labels in preclinical testing should be very precise indicating both the pain stimulus and the pain behavior.

Thermal (tail-flick, warm water tail-withdrawal, hot plate test, the Hargreaves test, immersion of the tail into cold water or lifting paws from a cold surface) and electrical stimuli are of short duration and are mainly used to study acute pain conditions. The use of thermal stimuli has numerous advantages, including translatability across species and reproducibility within- and between-subject responses. That is, at least, in part due to sharply defined sensory thresholds of thermal pain. Warm thermal stimulus tests such as tail-withdrawal test and hot plate test are good at predicting the analgesic activity of opioid analgesics from the group of agonist (morphinomimetics)[14, 17], and weaker when it comes to partial agonists (buprenorphine) or drugs from the group opioid receptor agonists / antagonists (pentazocine), as well as other less potent analgesics (non-steroidal anti-inflammatory drugs, paracetamol). As it is well known that drug screening of lower efficacy analgesics require lower intensity stimuli, it is suggested that new models of nociception should have the ability to provide a range of stimulus intensities and meet efficacy requirements in testing lower efficacy analgesics[54, 55] (Table 1). Electrical stimuli are applied to the tail, paw or dental pulp and responses such as twitching, escape behaviors, vocalizations, and biting are measured. Some studies have used an operant-based shock titration procedures[9]. Opioid analgesics have been shown to modify behavior after application of electrical stimuli, and their potency increases with the hierarchy of cortical involvement in each of the observed responses (motor responses < vocalization < vocalization after discharge)[20].

Mechanical stimulation can be used to induce both analgesia and hyperalgesia and often involve applying increasing pressure to an appendage (The Von Frey test, calibrated forceps, dolorimeter or Randall-Selitto analgesiometer). The animal responds by attempts to release appendage, visible struggle, vocalizations. A test that uses a mechanical stimulus such as paw pressure is suitable for testing the analgesic activity of nonsteroidal anti-inflammatory drugs (NSAIDs). The Von Frey test, allow measuring both mechanical hyperalgesia and allodynia[20].

Animal assays of CP became more prevalent in recent years and often involve the induction of inflammation and/or nerve injury.

Assays of tonic pain consider the application of different irritants (carrageenan, diluted formalin, acetic acid) in the rodent's skin, paw, muscle, joint, and visceral organs which produce nocifensive behaviors. Several of these substances can induce both an acute and chronic inflammatory response. The first phase in formalin test is an initial response resulting from nociceptor activation, and the second phase reflects inflammation[56]. The biphasic response allows researchers to study acute and longer-lasting tonic pain following a single noxious insult. For example, opioids can effectively suppress both acute and tonic pain, whereas NSAILs suppress only the tonic phase[57]. It is believed that among the models of acute pain, the formalin test has the greatest ability to predict the analgesic effect of new substances in humans. Beside mechanical and thermal hyperalgesia, the carrageenan injection can also enhance avoidance, spontaneous and guarding pain behaviors, as well as a reduction in the weight bearing force of an affected limb. The chemicals could also be given by intraperitoneal route or via injections into hollow organs to model visceral pain. Intraperitoneal administration of chemical agents induces stereotyped behaviors in the writhing test[58].

Complete Freund's adjuvant (CFA) consists of heat-killed Mycobacterium tuberculosis dissolved in mineral oils. CFA injections induce an immune-mediated chronic inflammation in tissues which modifies the behavior of animals during the conditioned place preference (CPP) and conditioned place avoidance (CPA) tests. CFA injection has been also used to model chronic inflammatory joint diseases such as in rheumatoid arthritis[59, 60].

Neuropathic pain models have been developed to mimic painful pathological conditions such as nerve trauma, nerve compression, low back pain or diabetic neuropathy[10]. The pain has been mainly modeled using the surgical intervention of a peripheral nerve such as sciatic nerve and its branches, or a spinal cord[61,62]. The most common models are nerve transection (NT), chronic constriction injury (CCI), partial sciatic ligation (PNL), spinal nerve ligation (SNL), spare nerve injury of the tibial and peroneal nerve (SNI), and inflammation.

Advances in animal pain research are being made in establishing disease models to represent cancer, osteoarthritis, post-operative, muscle and visceral pain, headache or migraine, etc[24,46]. Animal models of disease reflect the clinical condition the experimenter is trying to model, such as injection of CFA into the rat hind paw to study rheumatoid arthritis.

Assayes "batteries" combine operant behavior with a reflex-based model of nociception[63]. It allows characterization of behavioral selectivity (ratio of behaviorally disruptive to antinociceptive effects) of analgesic drugs. This may help in more effective selection of analgesic in accordance with individual treatment goals (Table 1).

Outcome measures

Outcome measures are designed to evaluate multiple parts of the pain experience and currently they are the most criticized aspects in pain re¬search. They can be generally categorized in evoked (tail-flick, paw withdrawal, vocalization following application of a stimulus (heat, pressure)) and non-evoked measures (spontaneous behaviors such as paw licking, flinching, biting, or altered weight-bearing)[50, 59, 64]. Measures of reflexive behaviors evaluate the behavioral response after noxious stimuli (heat, cold, mechanical, or electrical). Most of these behaviors are associated to spinal reflexes (limb/tail withdrawal), spino-bulbospinal reflexes (jumping and abdominal stretching) and innate behavior (vocalization, licking, scratching, biting, guarding) and can also be observed in decerebrated animals[10, 11, 50, 59]. Evoked measures are easy to perform and provide a quantifiable outcome. Evoked pain reflexes have clearly proven useful to investigate the underlying mechanisms of pain hypersensitivity (hyperalgesia and/or allodynia), identification of neurotransmitters, receptors and genes involved in pain behaviors; and better understanding of pain treatments[24]. Furthermore, the pharmacological action (efficacy, potency, duration of action) of traditional analgesics to reduce reflexive behavior in rodent models of acute and chronic pain have demonstrated reliable correspondence to human analgesia[53]. So, traditional models of reflexive behavior should not be abandoned or replaced; they should be refined (Table 1). The evoked measures do not consider the cognitive and emotional aspect of pain[10] and they are incapable to measure spontaneous pain states, which is the most common features of chronic pain[65]. On the other hand a great majority of clinical trials have measured spontaneous pain in human diseases. Therefore, new models that will help novel analgesics advancing from bench to bedside should incorporate cortical-dependent responses and assess multiple parts of the pain experience such as sensory and affective dimensions of pain and impact of pain on function and quality of life[24, 66](Table 1). They should provide much more manifestations of spontaneous pain relative to that induced in rodents by directed means. So an alteration in behavior of freely moving animals measured quantitatively in an observer-independent manner for a prolonged period of time is more preferable outcome measure than reflexive behavior[50]. Non-evoked measures have continuously demonstrated to be more concordant with clinical results than evoked measures (Table 1). An important limitation of these non-evoked paradigms is that they require significant learning with extensive training[5]. Some studies using operant measures of pain behavior have used paradigms such as condition place preference (CPP) and condition place avoidance (CPA)[52].

Non-evoked (operant) measures include: alterations in sleep, activity, eating and drinking, posture or gait, grooming, social interaction, and choice (preference, aversion) paradigms, pain-depressed behaviors, ultrasonic vocalization, facial expressions, etc.[50, 53] Using measures of function like wheel running or locomotor activity might be a useful measure to examine reduced physical activity, a common feature in clinical pain conditions. Nonetheless, evaluating chronic pain using only behavioral indicators has proven complicated. Rodents do not manifest pain in a consistent way and many of the proposed behaviors are not specific. Last years, electroencephalography, functional magnetic resonance imaging (fMRI) and positron emission tomography (PET) provided a great help in identifying specific areas of the brain that become activated during a painful experience and differentiating between mild and severe pain[5, 50]. In recent years, quantitative sensory testing (QST) which assess sensitivity of cutaneous and/or deep tissue, through a broad range of different sensations, is increasingly applied in both human and rodent studies[50].

Significant limitation of the standard animal models of pain is poor selectivity. It is related to the fact that these models are arranged such that the nonoccurrence of a response following a nociceptive stimulus is the primary evidence of antinociception. Because of that, these models are susceptible to false positives. For example, the effect of opioids or cannabinoids acting on motor activity in animals may easily be misinterpreted as pain-suppressing behavior[67, 68]. Another example of test with poor selectivity is the writing test (contractions of the abdominal wall and body contortions caused by intraperitoneal injection of dilute acetic acid) in mice, a very sensitive test used in the initial screening of new substances. This test may give false positive results with substances that do not have analgesic action (atropine, naloxone)[69]. In back translation, drugs identified as ineffective analgesics can be tested as negative controls, and can also be used to refine preclinical procedures in ways that increase selectivity against negative controls[5] (Table 1).

Novel approaches in animal pain researches should focus on the restoration of function (operant and other CNS-mediated behavior) under painful conditions[70]. Animals are active despite suffering a painful stimulus to earn a food reinforcer. The conflict between appetitive and nociceptive stimulation increase an animal's threshold of nociceptive stimulation. This will reduce the false-positive risk of confusing analgesia with behavioral disruption[71]. Also, in animal pain research a statistically significant increase in response thresholds is often used as a determinant of efficacy which leads to false positive results. In this way the efficacy bar is often set too low preclinically to allow successful translation to the clinical setting It has been shown that data analysis using effect size and NNT (Number Needed to Treat is the number of subject you need to treat to prevent one additional bad outcome) may provide better alignment with clinical data and allow successful translation to the clinical setting[24] (Table 1). In addition to schedule-controlled operant behavior mentioned before, the effects of nociceptive stimuli on naturally occurring behavior (nesting in mice, wheel running in rodents), were examined too[5, 72]. Evidence suggested that pretreatment with analgesics restore this CNS-mediated behavior which has been interrupted in the presence of a painful condition. Kandasamy[72] has been shown that following injection of Complete Freund's Adjuvant (CFA) into the rat hind paw CFA-induced depression of wheel running start to recover earlier than guarding the injected hind paw and CFA-induced allodynia assayed via Von Frey. This could be explained, at least partly, why humans often return to daily behaviors (work or socializing) before the symptoms of pain are completely relieved. It was suggested that these two models might be assaying different aspects of pain. These findings should encourage the development of models that allow for the investigation of novel treatments that restore daily function rather than simply eliminate hypersensitivity and also highlights the benefit of combining multiple models to investigate different aspects of pain[5, 20].

A hundreds of animal model/endpoint combinations have been developed and described in the literature, reflecting observations that pain phenotypes are mediated by distinct mechanisms. When an appropriate assays are determined for a potentially new analgesic, its antinociceptive profile can be successfully determined, ie. efficiency, potency, mechanism of action, site of action (topography), duration of action, development of tolerance, drug-drug interactions, etc.[24, 53]

Pain impacts nearly all aspects human/animal life, including function, activity, and quality of life and it is not always clear which of these is the main driving factor. As with animal models, clinical trial design should also consider basic research findings, and multiple outcomes measures (resting pain, movement pain, hyperalgesia, function and quality of life)[53]. IMMPACT (Initiative on Methods, Measurement and Pain Assessment in Clinical Trials), develops consensus reviews and recommendations for the measurement of pain treatment outcomes across numerous domains (pain, physical functioning, emotional functioning, participant ratings of improvement and satisfaction with treatment, symptoms and adverse events, participant disposition[73]. Translation in pain research should be a two-way process; clinical trials can be more appropriately designed to test a variety of outcomes based on the animal data and vice versa (Table 1). Discoveries in human medicine obtained through various approaches (brain imaging, genetic research) needs translation back into preclinical models to reach their full potential. As more pathology data on a clinical disease is discovered, modifying and adapting animal models and outcome measures will be more successful[53].

Ethical issues

It is generally accepted that animals are living beings that, like humans, suffer from pain and therefore deserve moral consideration and legal protection[23, 42]. There are a number of ethical issues that researchers should be aware of when conducting their pain research on animals. Many national and regional organizations (the European Union, the United States, Canada.) have provided comprehensive animal welfare guidelines and policies to ensure a high standard of animal care and use in pain research[74,75,76,77].

In experimental work, researchers must follow the recommendations of local Institutional Ethics Committees and ensure the safety, welfare and well-being of animals[42]. Behavioral models of pain require manipulation of experimental animals, and certain tests require their restraint (holding in wire or plexiglass holders that limit their mobility). If the animal is poor and under stress, its neurovegetative reactions will be overemphasized, which means that the observations of the researchers will not be valid. Also, stress is known to stimulate the release of endogenous opioids which can reduce pain perception (it can prolong latent times from pain stimulus administration to response). Since pain is registered on the basis of changes in the behavior of the experimental animal, it is necessary to take care of their well-being and eliminate stress as much as possible, both before and during the experimental procedure. Due to all this, it is necessary to train the animal to the experimental manipulation and procedure a few days before the experiment. This is important for both moral and scientific reasons[22].

Any intended infliction of pain in animals must have a clear and convincing statement as to the justification and relevance of the research (for clinical practice, for improving theoretical knowledge), as well as the competence and ability of researchers to monitor the pain in animals and make it the smallest possible. The more intense and prolonged the pain in animals, the greater is the justification and potential scientific contribution of research. In balancing between the pain and stress in animal and the scientific contribution of research, it must be assessed when the degree of pain is too great and then the animal is left out of the experiment and subjected to euthanasia. Pain in lower animal species cannot be considered less than in higher ones. The principle of equality of pain experiences between species allows for a difference only if a lower perception of a particular type of pain or a lesser presence of behavioral correlates of stress, fear, anxiety, and depression in humans is demonstrated[78, 79].

From the ethical point of view, it is easiest to research acute pain because, unlike chronic pain, animals always have the possibility of avoidance, or if they cannot avoid it, then there is a time limit for exposure to a painful stimulus. However, models of chronic pain in animals are necessary for the research of chronic pain conditions in humans. Since chronic animal pain cannot be avoided, these animals must be cared for and monitored with special care. Chronic perception of pain in animals is manifested, among other things, by changes in feeding, drinking water, sleeping, socializing, etc[79]. In order to reduce stress and pain in animals with chronic pain, analgesics, anesthetics, as well as sedatives should be used whenever possible, and euthanasia must be performed immediately after the end of the experiment.

There are several practical ethical recommendations for the examination of experimental pain in anaesthetized animals[22, 42].

(1) Planned experiments in which a non-anesthetized animal is exposed to pain must be considered by scientists as well as lay people before the start of research. It is necessary to show the potential benefit of applying this type of experiment to understand the mechanisms of pain and pain therapy. Also, during the research, the researcher must constantly justify the ethics of his research. (2) If possible, the researcher should try to apply the painful stimulus to himself; this applies to most non-invasive stimuli that cause acute pain. (3) In order to be able to assess the degree of pain,

the researcher must carefully monitor and evaluate any deviation of the experimental animal's behavior from normal behavior. For this purpose, physiological and behavioral parameters must be measured. This information must be entered in the manuscript. (4) In studies of acute and chronic pain in animals, measurements must be carried out to show with a certain degree of reliability that the animal is exposed to the least pain necessary for a given type of experiment. (5) An animal that is exposed to chronic pain for experimental purposes must be given painkillers, or provided with analgesics or pain relief as long as this does not affect the aim of the study. (6) Pain tests must not be performed on animals that are paralyzed due to the use of neuromuscular blockers without the use of general anesthetics or appropriate surgical procedures that would eliminate sensory consciousness. (7) Experiments in which animal pain is examined must be as short as possible and the number of animals used in the experiment as small as possible[78]. (8) Finally, when submitting manuscripts for publication or presenting results of their research on meetings, authors must confirm that they have adhered to ethical guidelines provided by national and regional organizations and that their research was approved by local Institutional Ethics Committees[22, 42, 79].

Conclusion

In humans, multiple emotional and cognitive factors influence the perception and experience of pain and this result in high inter-individual variability. However, pain in animals is mainly measured as a behavioral response to noxious stimuli, so that results obtained from animal studies are often more consistent, but not sufficiently predictive in the development of new drugs for the treatment of pain. At present, several recommendations are offered to overcome the current translational barriers in the development of novel pain drugs. The development of models for pain research in animals is moving more in the direction of fine modification (adjustment) of several standard procedures, but also completely new methods and approaches are being discovered. New models are being developed in parallel with the development of new drugs in the treatment of pain, and their goal is to better correlate with certain painful conditions in humans. Only time and practice will show whether they will result in an improvement in translational pain research and contribute to a better understanding of the mechanisms underlying pain generation, identifying new analgesic targets, and the development of new drugs. As ethical issues are concerned, it is generally accepted that animals suffer from pain, like humans, and for that reason deserve moral respect and legal protection. The researchers should adhere to ethical norms in research provided by national and regional ethical bodies and their research should be approved and supervised by local Institutional Ethics Committees.

Table 1. Current recommendations in translational pain research

- Novel animal models of pain should be characterized by translatability across species (human and nonhuman subjects) to best predict clinical utility.

- Translation in pain research should be bidirectional; clinical trials can be more appropriately designed to test a variety of outcomes based on the animal data and vice versa.

- Translational pain research must incorporate a determination of cross-species similarities and differences in the pain-related anatomy and physiology, including biopsychosocial processes too.

- Behavioral outcome measures in preclinical procedures should be homologous to clinically relevant outcomes in humans and both should include the measurement of variables that better reflect quality of life.

- In order to improve predictive validity of preclinical models, the author must take into consideration the gender, age, and comorbidities in the design of preclinical studies as well as phenotype characteristics on the clinical side.

- In recent years interest has grown in the development and implementation of large animal and/or naturally developing models of pain based on the assumption these will show greater predictive validity.

- Traditional animal models of pain have to be improved by considering supraspinally mediated contributions to pain (operant behavior) in addition to reflexive behavior.

- Traditional animal models of pain should not be abandoned or replaced. Most authors agree on the importance of continued refinement of currently used models as well as development of new more accurate models and outcome measures.

- In order to optimize both sensitivity and selectivity of preclinical testing, the use of combinations of preclinical procedures with complementary strengths and weaknesses is recommended.

- Models of nociception should have the ability to provide a range of stimulus intensities and meet efficacy requirements in testing lower efficacy analgesics

- Pain model data analysis using effect size and NNT may contribute to better alignment of preclinical and clinical data

- Drugs that failed clinical translation, can be back-translated preclinically as negative controls and used to refine preclinical procedures in ways that increase selectivity against negative controls.

References

1. Gereau RW 4th, Sluka KA, Maixner W, Savage SR, Price TJ, Murinson BB, Sullivan MD, Fillingim RB. A pain research agenda for the 21st century. J Pain 2014; 15:1203-1214.
2. Calati R, Laglaoui Bakhiyi C, Artero S, Ilgen M, Courtet P. The impact of physical pain on suicidal thoughts and behaviors: Meta-analyses. J Psychiatr Res 2015;71:16-32.
3. Cohen SP, Vase L, Hooten WM. Chronic pain: an update on burden, best practices, and new advances. Lancet. 2021;397:2082-2097.
4. Treede RD, Rief W, Barke A, Aziz Q, Bennett MI, Benoliel R, Cohen M, Evers S, Finnerup NB, First MB, Giamberardino MA, Kaasa S, Korwisi B, Kosek E, Lavand'homme P, Nicholas M, Perrot S, Scholz J, Schug S, Smith BH, Svensson P, Vlaeyen JWS, Wang SJ. Chronic pain as a symptom or a disease: the IASP Classification of Chronic Pain for the International Classification of Diseases (ICD-11). Pain 2019;160:19-27.
5. Negus SS. Addressing the Opioid Crisis: The Importance of Choosing Translational Endpoints in Analgesic Drug Discovery. Trends Pharmacol Sci 2018;39:327-330.
6. Vučković S. International Scientific Conference: Pain-Terrifying Lord of Life. Opioid Crisis Serbian Academy of Sciences and Arts, October, 22, Belgrade, 2018. Organizer: Faculty of Medicine, University of Belgrade, Editors: Mićić D., Stevanović P. 2018, 1959-1980.
7. Volkow ND, Collins FS. The Role of Science in the Opioid Crisis. N Engl J Med 2017;377:1798.
8. Millan MJ. The induction of pain: an integrative review. Prog Neurobiol 1999;57:1-164.
9. Le Bars D, Gozariu M, Cadden S.W. Animal models of nociception. Pharmacological Reviews 2001; 53:597-652.
10. Mogil JS. Animal models of pain: progress and challenges. Nat Rev Neurosci 2009;10:283-294.
11. King T, Porreca F. Preclinical assessment of pain: improving models in discovery research. Curr Top Behav Neurosci 2014;20:101-120.
12. Tomić MA, Vučković SM, Stepanović-Petrović RM, Ugrešić N, Prostran MŠ, Bošković B. The anti-hyperalgesic effects of carbamazepine and oxcarbazepine are attenuated by treatment with adenosine receptor antagonists. Pain 2004; 111:253-260.
13. Vučković SM, Tomić MA, Stepanović-Petrović RM, Ugrešić N, Prostran MS, Bošković B. The effects of alpha2-adrenoceptor agents on anti-hyperalgesic effects of carbamazepine and oxcarbazepine in a rat model of inflammatory pain. Pain 2006; 125:10-19.

14. Vučković S, Prostran M, Ivanović M, Došen-Mićović Lj, Todorović Z, Nešić Z, Stojanović R, Divac N, Miković Ž. Fentanyl analogs: structure-activity-relationship study. Curr Med Chem 2009;16:2468-2474.

15. Vujović KS, Vučković S, Vasović D, Medić B, Knežević N, Prostran M. Additive and antagonistic antinociceptive interactions between magnesium sulfate and ketamine in the rat formalin test. Acta Neurobiol Exp (Wars) 2017;77:137-146.

16. Srebro DP, Vučković SM, Dožić IS, Dožić BS, Savić Vujović KR, Milovanović AP, Karadžić BV, Prostran MS. Magnesium sulfate reduces formalin-induced orofacial pain in rats with normal magnesium serum levels. Pharmacological Reports 2018;70:81-86.

17. Jevtić II, Savić Vujović K, Srebro D, Vučković S, Ivanović MD, Kostić-Rajačić SV. Synthesis and pharmacological evaluation of novel cis and trans 3-substituted anilidopiperidines. Pharmacol Rep 2020;72:1069-1075.

18. Bennett GJ. Animal models of pain. In: Kruger L (ed). Methods in pain research. Boca Raton, Florida, CRC Press, 2001: 67-91.

19. Vučković S, Savić Vujović K, Ivanović M, Došen-Mićović Lj, Todorović Z, Vučetić Č, Prostran M, Prostran Milica. Neurotoxicity evaluation of fentanyl analogs in rats. Acta veterinaria 2012; 62:3-15.

20. Withey SL, Maguire DR, Kangas BD. Developing Improved Translational Models of Pain: A Role for the Behavioral Scientist. Perspect Behav Sci 2020;43:39-55.

21. Rittner and Dib-Hajj. 2022 Global Year for Translating Pain Knowledge to Practice: Welcome Letter from Co-Chairs. Available at: https://www.iasp-pain.org/publications/iasp-news/2022-global-year-for-translating-pain-knowledge-to-practice-welcome-letter-from-co-chairs/

22. National Research Council (US) Committee on Recognition and Alleviation of Pain in Laboratory Animals. Recognition and Alleviation of Pain in Laboratory Animals. Washington (DC): National Academies Press (US); 2009. 1, Pain in Research Animals: General Principles and Considerations. Available at: https://www.ncbi.nlm.nih.gov/books/NBK32655/

23. IASP Guidelines for the Use of Animals in Research. Available at: https://www.iasp-pain.org/resources/guidelines/iasp-guidelines-for-the-use-of-animals-in-research/ 2022

24. Whiteside GT, Pomonis JD, Kennedy JD. An industry perspective on the role and utility of animal models of pain in drug discovery. Neurosci Lett 2013;557 Pt A:65-72.

25. Schmidtko A, Lötsch J, Freynhagen R, Geisslinger G. Ziconotide for treatment of severe chronic pain. Lancet 2010;375:1569-1577.

26. Gimbel JS, Kivitz AJ, Bramson C, Nemeth MA, Keller DS, Brown MT, West CR, Verburg KM. Long-term safety and effectiveness of tanezumab as treatment for chronic low back pain. Pain 2014;155:1793-1801.

27. Quiding H, Jonzon B, Svensson O, Webster L, Reimfelt A, Karin A, Karlsten R, Segerdahl M. TRPV1 antagonistic analgesic effect: a randomized study of AZD1386 in pain after third molar extraction. Pain 2013;154:808-812.

28. Hill R. NK1 (substance P) receptor antagonists--why are they not analgesic in humans? Trends Pharmacol Sci 2000; 21:244–246.

29. Huggins JP, Smart TS, Langman S, Taylor L, Young T. An efficient randomised, placebo-controlled clinical trial with the irreversible fatty acid amide hydrolase-1 inhibitor PF-04457845, which modulates endocannabinoids but fails to induce effective analgesia in patients with pain due to osteoarthritis of the knee. Pain 2012; 153:1837–1846.

30. Yekkirala AS, Roberson DP, Bean BP, Woolf CJ. Breaking barriers to novel analgesic drug development. Nat Rev Drug Discov 2017;16:545-564.

31. Leenaars CHC, Kouwenaar C, Stafleu FR, Bleich A, Ritskes-Hoitinga M, De Vries RBM, Meijboom FLB. Animal to human translation: a systematic scoping review of reported concordance rates. J Transl Med 2019;1:223.

32. Hay M, Thomas DW, Craighead JL, Economides C, Rosenthal J. Clinical development success rates for investigational drugs. Nat Biotechnol 2014;32:40-51.

33. Arrowsmith J. A decade of change. Nat Rev Drug Discov 2012;11:17–18.

34. Yezierski RP. The Challenges of Translational Pain Research. Available at: https://www.iasp-pain.org/publications/relief-news/article/translational-pain-research/. 18 January 2022

35. Yezierski RP, Hansson P. Inflammatory and Neuropathic Pain From Bench to Bedside: What Went Wrong? J Pain 2018;19:571-588.

36. IASP Aim and Objectives of Global Year for Translating Pain Knowledge to Practice Available at: https://www.iasp-pain.org/publications/iasp-news/aim-and-objectives-of-global-year-for-translating-pain-knowledge-to-practice/.

37. Sneddon LU. Comparative Physiology of Nociception and Pain. Physiology (Bethesda). 2018;33:63-73.

38. Raja SN, Carr DB, Cohen M, Finnerup NB, Flor H, Gibson S, Keefe FJ, Mogil JS, Ringkamp M, Sluka KA, Song XJ, Stevens B, Sullivan MD, Tutelman PR, Ushida T, Vader K. The revised International Association for the Study of Pain definition of pain: concepts, challenges, and compromises. Pain 2020;161:1976-1982.

39. Zimmermann M. Behavioural investigations of pain in animals. In: Duncan IJH and Molony Y (eds). Assessing pain in farm animals. Office for Official Publications of the European Communities, Bruxelles, Belgium 1986:16-29.

40. Melzack R, Casey KL. Sensory, Motivational and Central control determinants of Chronic Pain: A New Conceptual Model. In: Dan R. Kenshalo (ed). The Skin Senses, Chapter: 20, Charles C Thomas, Illinois, USA, 2000; 423-439.

41. National Research Council (US) Committee on Recognition and Alleviation of Pain in Laboratory Animals. Recognition and Alleviation of Pain in Laboratory Animals. Washington (DC): National Academies Press (US); 2009. Available at: https://www.ncbi.nlm.nih.gov/books/NBK32658/ doi: 10.17226/12526

42. Vučković S, Savić Vujović K, Džoljić E Prostran M. Pain Research in animals-Ethical issues. In: Turza K, Todorovic Z, Prostran M (eds.). Bioethics and Pharmacology: Ethics in Preclinical and Clinical Drug Development, ISBN: 978-81-7895-579-7, Transworld Research Network, Kerala, India, 2012; 51-57.

43. Burma NE, Leduc-Pessah H, Fan CY, Trang T. Animal models of chronic pain: Advances and challenges for clinical translation. J Neurosci Res 2017;95:1242-1256.

44. Klinck MP, Mogil JS, Moreau M, Lascelles BDX, Flecknell PA, Poitte T, Troncy E. Translational pain assessment: could natural animal models be the missing link? Pain 2017;158:1633-1646.

45. Weyer AD, Zappia KJ, Garrison SR, O'Hara CL, Dodge AK, Stucky CL. Nociceptor Sensitization Depends on Age and Pain Chronicity(1,2,3).. eNeuro. 2016;3:ENEURO.0115-15.2015.

46. Berge OG. Predictive validity of behavioural animal models for chronic pain. Br J Pharmacol 2011;164:1195-1206.

47. Edwards RR, Dworkin RH, Turk DC, Angst MS, Dionne R, Freeman R, Hansson P, Haroutounian S, Arendt-Nielsen L, Attal N, Baron R, Brell J, Bujanover S, Burke LB, Carr D, Chappell AS, Cowan P, Etropolski M, Fillingim RB, Gewandter JS, Katz NP, Kopecky EA, Markman JD, Nomikos G, Porter L, Rappaport BA, Rice ASC, Scavone JM, Scholz J, Simon LS, Smith SM, Tobias J, Tockarshewsky T, Veasley C, Versavel M, Wasan AD, Wen W, Yarnitsky D. Patient phenotyping in clinical trials of chronic pain treatments: IMMPACT recommendations. Pain 2016;157:1851-1871.

48. Henze DA, Urban MO. Large Animal Models for Pain Therapeutic Development. In: Kruger L, Light AR, (eds). Translational Pain Research: From Mouse to Man. Boca Raton , Florida, CRC Press/Taylor & Francis; 2010. Chapter 17. Available at: https://www.ncbi.nlm.nih.gov/books/NBK57273/

49. Lascelles BDX, Brown DC, Maixner W, Mogil JS. Spontaneous painful disease in companion animals can facilitate the development of chronic pain therapies for humans. Osteoarthritis Cartilage 2018;26:175-183.

50. Herzberg D, Bustamante H. Animal models of chronic pain. Are naturally occurring diseases a potential model for translational research?. Austral Journal of Veterinary Sciences 2021;53:47-54.

51. Mercadante S, Porzio G, Gebbia V. Spinal analgesia for advanced cancer patients: an update. Crit Rev Oncol Hematol 2012;82:227-232.

52. Mogil JS. The Measurement of Pain in the Laboratory Rodent. Oxford University Press, Oxford, UK. 2018.

53. Gregory NS, Harris AL, Robinson CR, Dougherty PM, Fuchs PN, Sluka KA. An overview of animal models of pain: disease models and outcome measures. J Pain 2013;14:1255-1269.

54. Morgan D, Cook CD, Smith MA, Picker MJ. An examination of the interactions between the antinociceptive effects of morphine and various μ-opioids: The role of intrinsic efficacy and stimulus intensity. Anesthesia & Analgesia, 1999; 88:407–413.

55. Cook CD, Barrett AC, Roach EL, Bowman JR, Picker MJ. Sex-related differences in the antinociceptive effects of opioids: Importance of rat genotype, nociceptive stimulus intensity, and efficacy at the μ opioid receptor. Psychopharmacology 2000; 150:430–442.

56. Dubuisson D, Dennis SG. The formalin test: a quantitative study of the analgesic effects of morphine, meperidine, and brain stem stimulation in rats and cats. Pain 1977;4(2):161-174.

57. Jourdan D, Ardid D, Bardin L, Bardin M, Neuzeret D, Lanphouthacoul L, Eschalier A. A new automated method of pain scoring in the formalin test in rats. Pain 1997 ;71:265-270.

58. Niemegeers CJ, Van Bruggen JA, Janssen PA. Suprofen, a potent antagonist of acetic acid-induced writhing in rats. Arzneimittelforschung. 1975;25:1505-1509.

59. Gregory NS, Harris AL, Robinson CR, Dougherty PM, Fuchs PN, Sluka KA. An overview of animal models of pain: disease models and outcome measures. J Pain Symp Manage 2013;14:1255–1269.

60. Parvathy SS, Masocha W. Gait analysis of C57BL/6 mice with complete Freund's adjuvant-induced arthritis using the CatWalk system. BMC Musculoskelet Disord 2013;14:14.

61. Jaggi AS, Jain V, Singh N. Animal models of neuropathic pain. Fundam Clin Pharmacol 2011;25:1-28.

62. Goel SA, Varghese V, Demir T. Animal models of spinal injury for studying back pain and SCI. J Clin Orthop Trauma 2020;11:816-821.

63. Withey SL, Paronis CA, Bergman J. Concurrent Assessment of the Antinociceptive and Behaviorally Disruptive Effects of Opioids in Squirrel Monkeys. J Pain 2018;19:728-740.

64. Krug HE, Dorman C, Blanshan N, Frizelle S, Mahowald M. Spontaneous and Evoked Measures of Pain in Murine Models of Monoarticular Knee Pain. J Vis Exp 2019;144:10.3791/59024.

65. Backonja MM, Stacey B. Neuropathic pain symptoms relative to overall pain rating. J Pain 2004;5:491-497.

66. Vierck CJ, Hansson PT, Yezierski RP. Clinical and pre-clinical pain assessment: are we measuring the same thing? Pain 2008;135:7-10.

67. Meng ID, Manning BH, Martin WJ, Fields HL. An analgesia circuit activated by cannabinoids. Nature 1998;395:381–383.

68. Vučković S, Srebro D, Vujović KS, Vučetić Č, Prostran M. Cannabinoids and Pain: New Insights from Old Molecules. Front Pharmacol 2018;9:1259.

69. Ness TJ. Models of Visceral Nociception. ILAR J 1999;43:119-128.

70. Kangas BD, Bergman J. Operant nociception in nonhuman primates. Pain 2014;155:1821-1828.

71. Rohrs EL, Kloefkorn HE, Lakes EH, Jacobs BY, Neubert JK, Caudle RM, Allen KD. A novel operant-based behavioral assay of mechanical allodynia in the orofacial region of rats. J Neurosci Methods 2015;248:1-6.

72. Kandasamy R, Calsbeek JJ, Morgan MM. Home cage wheel running is an objective and clinically relevant method to assess inflammatory pain in male and female rats. J Neurosci Methods 2016;263:115-22.

73. Dworkin RH, Turk DC, Farrar JT, Haythornthwaite JA, Jensen MP, Katz NP, Kerns RD, Stucki G, Allen RR, Bellamy N, Carr DB, Chandler J, Cowan P, Dionne R, Galer BS, Hertz S, Jadad AR, Kramer LD, Manning DC, Martin S, McCormick CG, McDermott MP, McGrath P, Quessy S, Rappaport BA, Robbins W, Robinson JP, Rothman M, Royal MA, Simon L, Stauffer JW, Stein W, Tollett J, Wernicke J, Witter J. Core outcome measures for chronic pain clinical trials:IMMPACT recommendations. Pain 2005; 113:9–19.

74. Directive 86/609/EEC, European Economic Community, November 24, 1986

75. Canadian Council on Animal Care. Guide to the Care and Use of Experimental Animals Vol. 1. 2d ed. Ontario, Canada: CCAC, 1993. WWW: http://www.ccac.ca/

76. National Academy of Sciences. Guide for the Care and Use of Laboratory Animals, 7th ed. Washington, D.C.: National Research Council, Institute for Laboratory Animal Research, NAS, 1996. WWW: http://www.nap.edu/catalog/5140.html

77. National Institutes of Health. OPRR Public Health Service Policy on Humane Care and Use of Laboratory Animals. Rockville, MD: NIH/Office for Protection from Research Risks, 1996. WWW: http://grants.nih.gov/grants/olaw/olaw.htm

78. Tannenbaum J. Animal Models of Pain. Ethics and Pain Research in Animals. ILAR Journal 1999; 40(3).

79. Allen C. Animal Pain. Noûs 2004; 38:617

USING ANIMALS IN BRAIN RESEARCH

Željka Stanojević and Ivanka Marković

University of Belgrade, School of Medicine, Belgrade, Serbia

Abstract

The use of animals in biomedical and behavioral research has greatly increased scientific knowledge and has had enormous benefits for human health. Every Nobel Prize in Medicine awarded in the last three decades was dependent on data from animal models. Overall, 83% of the Nobel Prizes awarded for outstanding contributions to medicine have involved animal research since the program was founded in 1901. A particular challenge is to find a balance between the need for research on animal models and the welfare of experimental animals. Here, we discussed about Russell and Burch set of 3Rs and its application in experimental medicine, usefulness and beneficial information in animal brain research, their alternatives like *in silico*, *in vitro* and *ex vivo* alternatives with its advantages and disadvantages. Also, we discussed moral concerns about use of nonhuman primates and the regulations that regulate it.

Keywords: bioethics, animal models, alternative models, nonhuman primate

1. Use of animals in brain research

The use of animals in biomedical and behavioral research has greatly increased scientific knowledge and has had enormous benefits for human health.

Every Nobel Prize in Medicine awarded in the last three decades was dependent on data from animal models. Overall, 83% of the Nobel Prizes awarded for outstanding contributions to medicine have involved animal research since the program was founded in 1901, more than 100 years ago (https://fbresearch.org/medical-advances/nobel-prizes/).

Along the same lines, animal research and testing were needed for every prescription medicine available today. The U.S. Food and Drug Administration (FDA) requires animal testing to ensure the safety of many drugs and devices.

The axiom "before testing a new treatment in man, test it first in animals if possible" has been part of drug development for the past 50 years or so. Testing in animal models is believed to increase the chances of identifying drugs that are sufficiently promising to justify the effort and expense of further clinical development (1). Virtually every major medical advancement of the last century has depended upon research with animals. Animals have served as surrogates in the investigation of human diseases and have yielded valuable data in the process of discovering new ways to treat, cure or prevent them. From immunizations to cancer therapy, our ability to manage the health of animals has also improved because of animal research and the application of medical breakthroughs in veterinary medicine.

The use of animals in scientific research has been a controversial issue for well over a hundred years. The basic problem can be stated quite simply: Research with animals has saved human lives, lessened human suffering, and advanced scientific understanding, yet that same research can cause pain and distress for the animals involved and usually results in their death. It is hardly surprising that animal experimentation raises complex questions and generates strong emotions.

Perhaps more than in any other field of biomedical research, it is essential to use animals to understand the functions of the brain, both in basic research and drug testing. Brain disorders can include any problem with the brain and spinal cord, including mental health and sensory disorders. However, one of the greatest challenges in neuroscience research is tackling neurodegenerative diseases such as dementia and Parkinson's disease, which currently affect tens of millions of people across Europe. As the proportion of the elderly in Europe increases, it is vital that the most effective methods of research are used to combat this challenge (2).

To respond to these concerns, the Center for Neuroscience and Society (CNS) and the Center for the Interaction of Animals and Society at the University of Pennsylvania hosted a workshop on the "Neuroethics of

Animal Research" in Philadelphia, Pennsylvania, in June 2016. This workshop has three recommendations: 1) Develop Guidelines for Evaluating Animal Models of Psychiatric Conditions, 2) Allocate Greater Funding to in Vitro Neurological Models, 3) Conduct Welfare Research on the External Control of Animals (3).

At present, society's best hope of finding drugs and other treatments for diseases of the brain rely on research using animals. While non-animal methods of study have made progress in some fields of biomedical research, their use in neuroscience remains extremely limited due to the complex and interconnected structure of the brain. In most cases, a living and behaving organism remains the only viable model to study the brain in action.

Which animals are used in brain research?

Mice and rats – "Rodents are the dominant mammalian animal species used in neuroscience research", said Bill Yates, professor of otolaryngology and neuroscience at the University of Pittsburgh, but the Animal Welfare Act excludes mice and rats, so the exact number used in the United States is not available (4). The number of higher animals used is known because the U.S. Department of Agriculture (USDA) requires research institutions to submit an annual report of the number of animals used. The use of most animal species tracked by the USDA has declined over the past decades (The most commonly used animal species in neuroscience research are mice and rats, as the complexity of their brains is similar enough to humans to give a good overview of brain processes. Mice can also be genetically modified with relative ease, meaning that researchers can look at the effect of individual genes on the way disease progresses.

As mice and rats have a shorter life span than other mammals, it is possible to use them to study diseases over a longer period of time or in aging animals. Mice and rats can also be used to study the effects of additional conditions (comorbidities), such as obesity or diabetes, on neurological diseases.

Transgenic mouse models

The most significant contributor to the increased use of rodents in biomedical research has been the development of transgenic mouse models. In the late 1980s, Capecchi, Evans and Smithies developed principles for introducing specific gene modifications in mice by the use of embryonic stem cells leading to the development of the first knockout mouse (5). Today, human genes can be inserted into a mouse or overexpress a particular gene. Through breeding, it is possible to obtain a line of animals that expresses a new phenotype (6). Most procedures now are done using transgenic animals. Data suggest that transgenic mice likely account for two-thirds or more of the mice, and more than half of the mammals used in the biomedical research.

Use of transgenic animals has allowed neuroscientists to decipher the function of particular genes and to create disease models. Knockout models have been used in the study of Alzheimer's disease, for example, and have been critical in understanding the neural basis of learning and memory (7). In addition, some genetic diseases have different phenotypes in mice and humans. For example, transgenic models of Parkinson's disease often do not exhibit the same neural degeneration observed in humans (8,9,10).

Likewise, compensation for gene manipulation during development can lead to false conclusions about the role of particular genes.

Zebrafish - Zebrafish are also becoming more widely used for neuroscience research. Together with mammals and popular invertebrate model species (Figure 1), both larval and adult zebrafish are extensively used in central nervous system (CNS) research and targeting various brain disorders (11,12). Observational studies, combined with the ability to see the molecular changes in the brain and firing neurons, are ongoing to understand how they react to stress (13) and depression (14) and used to test potential treatments or highlight new mechanisms to study in humans.

Non-human primates - Having such a close genetic relationship to humans, non-human primates (NHPs) are one of the most valuable animal models used in research. Less than one percent of the animals used in research in the EU are NHPs, however their impact in providing the most reliable information for what is happening, or what is going to happen, in humans cannot be underestimated.

While there are understandable ethical worries about using NHPs, they continue to be an important model for studying the function of the brain due to the similarity in structure and composition. As NHPs have very similar brains to humans, we can gain reliable information as to what might be happening in a human one. Much of what we know today about complex behavior and emotion, vision, and higher cognitive function was gained from the studies in NHPs.

Still more options are on the horizon: the 2015 NIH decision to eliminate most NHPs research on the grounds that such research was unnecessary is evidence that the current process for approving research is not robust enough to rule out animal models that are not justified by a true harm-benefit analysis (15).

This conclusion highlights an important ethical point in which it is discussed in the following segment.

Can we replace animals with alternatives?

Russell and Burch set of 3Rs (Replacement, Reduction, and Refinement)

Conflicting feelings about conducting experiments in animals have existed for more than 4 centuries (16). The 'father of physiology,' Claude Bernard (1813–1878) recognized this polarization in his statement that "the science of life is a superb and dazzlingly lighted hall which may be reached only by passing through a long and ghastly kitchen."(17). In fact, Dr. Bernard's wife, Marie Francoise Martin, established the first antivivisection society in France (17). The conflict, concern, and debate continue into the present time.

A few examples give an indication of the breadth and variety of these contributions: a) Animals have been used to study cardiovascular function and disease since the early 1600s. Heart-lung machines, which have made open-heart surgery possible, were developed with animals before being used with humans (18) b) Studies of the biology of transplantation in animals have made it possible to transfer organs between people (19), c) Animal research shed light on the nature of polio and has helped to nearly eliminate the disease (20,22), d) Many clinically useful methodologies were first tested on animals before being used with humans. Examples include computed axial tomography (CAT) scans and magnetic resonance imaging (MRI) (23), e) Animal studies have been essential in probing the functions of the brain in health and disease (23). Investigators have used animals to understand movement (and the movement dysfunctions caused by such diseases as epilepsy and multiple sclerosis), vision, memory (including the severe memory loss that occurs in 5 percent of persons over the age of 65), drug addiction, nerve cell regeneration, learning, and pain disease (22).

One major step forward was taken with Russell and Burch's seminal book on animal research. Russell and Burch developed what is known as the 3Rs approach more than 50 years ago with the intention of balancing the advancement of knowledge in science with respect for the lives and experiences of animals.

From the standpoint of Russell and Burch there are, after all, 3Rs principles (23). These principles assert that whenever possible, animal models should be replaced with alternative methods; the number of animals used in experiments should be *reduced* to a minimum; and their suffering should, whenever possible, be ameliorated, e.g., through human endpoints, less invasive procedures, and the use of anaesthesia (*refinement*). Gradually, however, they have become established as essential considerations when animals are used in research. They have influenced new legislation aimed at controlling the use of experimental animals, and in the United Kingdom, they have become formally incorporated into the Animal (Scientific) Procedures Act (24). The three principles, *Replacement, Reduction and Refinement*, have also proven to be an area of common ground for research workers who use animals, and those who oppose their use. Scientists, who accept the need to use animals in some experiments, would also agree that it would be preferable not to use animals. If animals were to be used, as few as possible should be used and they should experience a minimum of pain or distress. Many of those who oppose animal experimentation would also agree that until animal experimentation is stopped, Russell and Burch's 3Rs provide a means to improve animal welfare. It has also been recognised that adoption of the 3Rs can improve the quality of science. Appropriately designed experiments, which minimise variation, provide standardised optimum conditions of animal care and minimise unnecessary stress or pain, often yield better and more reliable data.

Despite the progress made as a result of the attention to these principles, several major problems have been identified. When replacing animals with alternative methods, it has often proven difficult to formally validate the alternative. This has presented a particular problem in regulatory toxicology, especially when combined with the labyrinthine processes of the various regulatory authorities (25). The principle of Reduction would appear less contentious, but its application has highlighted the difficulties of providing appropriate expert statistical advice, especially in academic research facilities (26). In some instances, concern to implement reduction strategies can result in the use of too few animals, which leads to inconclusive results, and wasteful experiments. It is in the area of Refinement, however, that major problems have arisen. Much of our judgment of what represents Refinement is based on little more than common sense. We make assumptions about animals and their feelings that often have little scientific basis (27). In many instances we may be correct, but these assumptions may become incorporated into institutional or national policies, without any attempt to verify them.

At one end of the spectrum, Thomas Hartung, who successfully replaced animal experiments with cell cultures in his graduate research, is now director of the Center for Alternatives to Animal Testing, or CAAT, at the Johns Hopkins Bloomberg School of Public Health. He believes the biggest step researchers can take toward reducing animal experimentation is a change in mindset; animal models have been useful for so long, it's hard to believe that there are better models out there, even if the data exists (28).

There is, after all, many *in silico, in vitro* and *ex vivo* alternatives are also available to researchers in pharmaceutical companies such as physico-chemical techniques, microbiological systems, cell/tissue cultures, isolated organs, 3D tissue cultures, computer or mathematical analysis, epidemiological surveys, plant analysis, physiology based pharmacokinetic modeling, endpoint assays, microarray technology, human clinical trials utilizing microdosing, epidemiological studies, "omics" technologies, stem cells, DNA chips, microfluidics chips, human tissues and voluntary donated organs, and a lot of new imaging technologies (29).

The term "alternatives" encompasses a range of options. In the research community, an alternative has been defined to mean reducing the number of animals used, refining experimental designs to lessen any pain or distress in animals, or replacing animals with other organisms or techniques. An alternative may therefore still involve the use of animals, but it might mean using fewer animals or using them in different ways.

To help clarifying this issue it is necessary to understand that most researchers generally hold that non-animal experiments are adjuncts rather than alternatives to animal experiments. Studies that do not use animals can produce valuable information, but they cannot completely replace the information gained from animal experiments. Only animals can demonstrate the effects of a disease, injury, treatment, or preventive measure on a complex organism. For example, some aspects of arthritis cannot be studied in tissue culture cells without bones or joints (30).

In any case, animal-free alternative methods in neuroscience research are also an important way to study the brain. Organoid mini-brains enable researchers to test the effects of some drugs, and can provide an insight into how the brain develops and, combined with computer simulations, they can offer a snapshot of how cells act and develop at any given time. The EU Commission Joint Research Centre has published a collection of over 550 non-animal models used for studying neurodegenerative diseases.

In addition to "traditional" alternatives to animal experimentation (like tissue cultures), several innovations have been recently introduced with the objective to retrench animal experimentation. They include organs on chips, human-derived three-dimensional tissue models, human blood derivates, microdosing, and computer modelling. However, these alternatives can only reduce research dependence on animals (through replacement and reduction) by complementing animal research.

However, these cell-based methods are often of limited use, consisting of only one cell type, or grown in isolation of other tissues, the immune system and blood supply. They also lack the interactions with other organs of the body, particularly important now that we are beginning to understand more about the relationship of the brain with the gut, heart, and lungs.

Discussion of these problems should not detract from the very significant progress that has been made in the 40 or so years since Russell and Burch set out their guiding principles.

While we wait for alternative methods of study to emerge it is therefore essential to continue to develop better animal models and ensure the highest standards of animal welfare.

2. Usefulness and beneficial information in animal brain research

This segment should be viewed within the context of the vast improvements in human health and understanding that have occurred in the past 150 years.

Despite many advances and the projected results that will come through the use of animals, some individuals question the value of using animal models to study human disease, contending that the knowledge thus gained is insufficiently applicable to humans. Although experiments performed on humans would provide the most relevant information, it is not possible, by commonly accepted ethical and moral standards, or by law, to perform most experiments on humans initially. It is true that not every experiment using animals yields immediate and practical results, but the advances that will be described in this chapter provide evidence that this means of research has contributed enormously to the well-being of humankind.

There are many animal models in neuroscience, useful or less useful: some that shed light on the pathogenesis of the disease, some served to test drugs, while some served only to analyze clinical manifestations. The most vivid example of the advantages of animal models, as well as limitations in neuroscience, is experimental autoimmune encephalomyelitis (EAE), an animal model of multiple sclerosis.

Although the cause and pathogenesis of multiple sclerosis (MS) are unknown, current prevailing hypothesis states that MS represents an autoimmune disorder directed against nervous system antigens (31,32). The basic concept proposes that exposure to environmental pathogens activates autoreactive T cells that recognize central nervous system (CNS) autoantigens, leading to inflammation and demyelination (33;34).

Since the initial experiments by Rivers, the stage was set for the use of experimental animal models to study CNS inflammation and demyelination (35). Over the last 30 years, the number of EAE-cited publications in English has quadrupled; a Medline search identifies a total of 678 articles on EAE between the years 1970 and 1980, 1,860 articles between 1990 and 2000, and approximately 1,600 publications since 2001. Besides the utilization of EAE to study MS, it has also been harnessed for developing therapeutic strategies for MS (36,37,38). Indeed, the majority of the current therapies being planned for phase II and III trials in MS were first examined in EAE. Furthermore, immunotherapies are successfully used as disease-modifying therapies (DMTs) in MS (39, 40). EAE has been extremely valuable in determining the efficacy of DMTs. Several DMTs are now approved for the treatment of patients with MS and most of them have previously been tested in EAE: three formulations of IFN-β, glatiramer acetate (GA), mitoxantrone, natalizumab, and three oral drugs (i.e., fingolimod, teriflunomide, and dimethyl fumarate) (41). Thus, EAE has become a central player in the arena of MS. Is it indeed a suitable and relevant research tool for MS? It has improved our understanding of acute inflammatory demyelinating syndromes, advanced our knowledge of the genetic susceptibility to autoimmunity, and helped uncover mechanisms of lymphocyte trafficking and the role of blood–brain barrier in CNS inflammation.

At the other end of the spectrum, there is another point of view: Steinman and Zamvil believe that the EAE model has lots of limitations and they state that encouragement of industrial scientists to adopt the EAE model for drug screening is analogous to Michael Jordan suggesting that all citizens should attain fitness by learning to soar through the air and execute elaborate slam-dunks. Such performances look easy in the hands of virtuosos but cannot be reproduced without years of practice, which the novice does not see (42).

Indeed, EAE pathogenesis is exceedingly complex and amelioration of EAE is not a strong predictor of success in treating MS. EAE is not a disease but a scientific tool that comes in many forms, each of which can be used to address specific questions. The authors' list of EAE models and variants shows how difficult it would be to use this 'model' (actually dozens of experimental paradigms) for drug screening.

After all, it looks like it would be interesting to ask the question of how one could approach the disease if animal models were unavailable, and the only recourse would be to examine the clues offered by our patients and from relevant genetic, imaging, and epidemiological studies in humans? We believe that the current available pathological, as well as radiological data, would argue favorably in examining issues outside of the "autoimmune hypothesis" as central elements in the disease process.

3. Moral concerns about global increase in the use of non-human primates

Since the mid twentieth century, NHPs have been widely used in laboratory research, mostly in biomedical research (43), and mostly in the cognitive sciences (44). In recent years, due to public pressure and legislation, the number of NHPs used in biomedical research was significantly reduced in the European Union and the United States (45, 46), but has increased dramatically in some other countries, China in particular (47). Some of the very first "humanlike" capabilities that attracted considerable scientific interest were the discoveries that NHPs build and use tools (48,49), solve new problems, and develop and pass on cultural behaviour (50,51, 52,53). All NHPs that have been studied to date, from Rhesus NHPs to chimpanzees and gorillas, have also been shown to have distinct personalities with complex behavioural patterns, as occurs with humans (54). Furthermore, all NHPs establish strong social bonds (55), and most live in complex societies (56). Like humans, NHPs experience and display emotions (57), strong mother–infant and other familial bonds and are capable of experiencing empathy and behaving sympathetically (58). Because of their anatomical and physiological similarity to humans, as well as such cognitive, behavioural, and social similarities, NHPs have been portrayed as ideal animal models for some biomedical and cognitive research.

It has been pointed out that the similarities between humans and NHPs are the main ethical obstacle regarding the laboratory confinement and use of NHPs (59). This is indeed a controversial issue within the scientific community and for the general public (60), but the recognition of significant ethical concerns that need to be addressed is nearly universal. Whilst public health advancement might be a justifiable goal, from a utilitarian standpoint, the pursuit of biomedical NHP research (that might provide only modest benefit) might not be justifiable. From a deontological point of view, the 3Rs are largely irrelevant, since they do not prevent the research subjects from being used as means to an end. Additionally, the 3Rs do not comply with principles of autonomy or justice, which are crucial within the deontological approach prescribed by Beauchamp and Childress (61).

Moral concerns raised by the use of NHPs in intrusive or invasive research result from their sentience, consciousness and affective states. In those aspects, NHPs are very similar to humans, which make it reasonable to give them similar protection to that afforded to human subjects. However, they are very different in other aspects, so they are not necessarily good models for human biology.

Regarding clinical research, the regulations and mechanisms like The International Ethical Guidelines for Biomedical Research Involving Human Subjects and Council for International Organizations of Medical Sciences (CIOMS) seem to be effective in solving most of the ethical challenges (62) but that is not the case with invasive animal experiments, particularly those using NHPs (63). The main reason for these inconsistencies seems to be the use of different frameworks to evaluate and guide research with humans and NHPs (64). Guidelines and legislation that regulate human research rely on mostly deontological principles, while those that regulate animal research rely on utilitarianism.

Deontological ethics

The philosopher Immanuel Kant (65) introduced the concept of deontological ethics; hence, deontological ethics is also called Kantian deontology. Being a devout Christian, Kant grounded his duty-based ethical principles in terms of universal moral obligations. Moreover, thinking that each human being has an inherent value, Kant thinks that the autonomy, dignity, and respect concerning each individual should be emphasized. Ross (66) modified Kant's deontology, allowing a plurality of duty-based ethical principles,such as doing no harm, promise keeping, etc. In contrast to utilitarian principles, deontology principles refer to the ethics of duty, in which no harm is allowed, even if it may lead to positive consequences (67). Hence, decisions made based on deontological ethics may be appropriate for an individual even though those decisions may not lead to good outcomes for society as a whole (67).

Deontological ethics emphasizes the value of being a human being, underlining the principles of respect for autonomy, beneficence, non-maleficence, and justice (68). Therefore, deontological ethics can help medical care professionals further understand the four principles, regarding respect for autonomy, non-maleficence, and justice as the principles of humanity values and beneficence as a principle of maximizing human happiness and relieving suffering (68).

As stated earlier, although the philosopher most associated with deontology, Immanuel Kant, has thought that humans hold a different place in the world order than animals, there are other deontologists who strongly believe it is wrong to treat animals as a means to an end. This stance means non-human animals should be viewed in the same way Kant viewed human animals in his second formulation of his Categorical Imperative, the "Formula of the End-in-Itself."(69). To paraphrase, deontological beliefs require that you do not treat others as simply a means to an end but as the end. This is often distilled down to "do not use others". If a person truly sees no difference between human and non-human animals, application of Kant's principles would mean that very little or perhaps no research would be acceptable. So a strong deontological view like Regan's does not provide much guidance about how to specifically apply the 3Rs.

Other deontological views are more closely aligned with the 3Rs. Tannenbaum (70) has an essentially deontological theory stating that animals have at least one right: the right for their interests to be taken into consideration. In this view, the capacities of animals mentioned above are relevant in researchers 'decisions about the design of experiments.

Utilitarianism

Although some deontological views are compatible with the 3Rs approach, the view most commonly associated with the 3Rs is utilitarianism (71). According to utilitarianism, animal research can be justified if the benefits to humans and animals outweigh the harms caused to animals during research.

Along that line, the founder of modern utilitarian ethics, Jeremy Bentham, introduced in An Introduction to the Principles of Morals and Legislation (72) the principle of utility for the evaluation of appropriate actions. The rightness or wrongness of a selected action is decided according to whether the action would maximize a positive outcome, that is, whether the action would bring less pain and more pleasure to the most people. Bentham (72) quantifies the amount of pain and pleasure created from actions in a moral utilitarian calculus that examines the rightness or wrongness of the selected actions in terms of seven factors: intensity, duration, certainty, propinquity or remoteness, fecundity, purity, and extent (73). Utilitarian ethics is a version of consequentialist ethical theories. Although there are different varieties of utilitarian ethical principles, the basic idea of these principles is based on Bentham's theory: maximize utility and prioritize happiness.

It seems that the majority of utilitarian views are either hedonistic, emphasizing conscious experiences like pleasures and pains as benefits and harms, or preference–satisfaction views, which consider desire satisfaction and desire frustration as benefits and harms, respectively. Because utilitarian theories claim that right actions are those that maximize benefits and minimize harms, there is a strong incentive to reduce harms as much as possible.

In other words, deontology is patient-centered, whereas utilitarianism is society-centered. Although these approaches contradict each other, each of them has their own substantiating advantages and disadvantages in medical practice. Over the years, a trend has been observed from deontological practice to utilitarian approach leading to frustration and discontentment. Health care system and practitioners need to balance both these ethical arms to bring congruity in medical practice.

Putting it all together: The major schools of philosophy of utilitarianism and deontology, although different, are not necessarily incompatible. For example, people may hold a view of rights for people and utilitarianism for animals. Therefore, we might think that strong deontological considerations should protect humans from being subjected to harmful experiments against their will, but nevertheless believe that it is permissible to perform research on animals if such research results or is likely to result in gains that outweigh the harms.

4. Genetic modification of NHPs, and their development as models of human brain disorders

The application of genetic engineering technologies, from basic research in animal models to clinical applications in cancer therapy, has revolutionized biomedical research, including neuroscience research. Until recently, the use of these technologies has been limited mostly to rodents and other lower model organisms. While studies using a variety of animal model systems have dramatically enriched our knowledge of molecular, cellular, and systems neuroscience, there has been limited impact on our understanding of higher human brain function, such as emotional states, cognitive function, and social interaction, partially due to structural and functional differences between rodent and human brains.

The recently developed, highly efficient new genome-editing technologies, such as the CRISPR/Cas system, now make it feasible to expand genetic engineering to many other species (74), thus opening the door for generating genetically modified NHP models for basic neuroscience and brain disorders research. Such models are urgently needed if we are to make progress in understanding higher brain function and related disorders in humans.

Genetic engineering can be performed at multiple developmental stages and with a variety of approaches. Germline manipulations are likely the most valuable in modeling human genetic mutations. However, due to the current low efficiency in generating large numbers of mutant founder animals, combined with the long waiting time for sexually mature NHPs to produce offspring, genetic engineering in somatic cells offers useful approaches.

The ability to genetically modify the genome in NHPs to generate cell type-specific tools and disease models has the potential to transform study of higher brain function and dramatically facilitate the development of effective treatment for human brain disorders.

Responsible use of genetically modified NHP models

While it is clear that genetically modified NHP models have great potential for utilization in translational neuroscience, these unique animal models must be well-justified for each application. Each project must meet at least one of the following criteria: a) Have a clear scientific understanding and justification that an NHP model is the best way to address an important question, b) Have clear evidence of failure using other models to address important basic or translational questions, c) Have a history of success with other models, suggesting that an important basic or translational question might be answered (75).

Ethical considerations

Responsible use, as described above, also implies ethical use. Ethical issues are important in the creation and use of genetically modified NHPs for research purposes, as they have long been with all research with non-human animals. Research with non-human animals is justified when one can reasonably expect sufficient benefits to humans or benefits to science—the latter often having unforeseen benefits for humans and potentially other animals—to justify the risks or harms to the nonhuman animals (76).

It should be noted that the "three Rs"—replace, reduce, and refine—which are the guiding principles in research with nonhuman animals (described above), will need to be applied carefully in this context of genetically modified NHPs. The possibilities opened by genetic modification of NHPs may in some specific instances actually lead to more, not less, use of

NHPs in research. Any short-term increase needs to be justified by greater potential benefits to humans or science, but the goals of minimizing the numbers of animals used and the risks and harms to which they are exposed will still apply.

References

1. Sandercock P, Roberts I. Systematic reviews of animal experiments. Lancet. 2002;360:586.

2. Roberts I, Kwan I, Evans P, Haig S. Does animal experimentation inform human health care? Observations from a systematic review of international animal experiments on fluid resuscitation. BMJ 2002;324:474-6.

3. Adam J Shriver, Tyler M John. Neuroethics and Animals: Report and Recommendations from the University of Pennsylvania Animal Research Neuroethics Workshop. ILAR J. 2021;60:424-433.

4. International Animal Research Regulations: Impact on Neuroscience Research: Workshop Summary.

5. Torbjørn Hansen. The Nobel Prize in physiology or medicine 2007. Scand J Immunol. 2007;66:603.

6. Gregory Prelich. Gene Overexpression: Uses, Mechanisms, and Interpretation. Genetics. 2012; 190: 841–854.

7. Hsiao K, Chapman P, Nilsen S, et al. Correlative deficits, Abeta elevation, and amyloid plaques in transgenic mice. Science 1996;274:99-102.

8. Ted M. Dawson, Han Seok Ko and Valina L. Dawson. Genetic Animal Models of Parkinson's Disease. Neuron. 2010; 66: 646–661.

9. J. A. Potashkin, S. R. Blume, and N. K. Runkle. Limitations of Animal Models of Parkinson's Disease. Parkinson's Dis. 2011; 2011: 658083.

10. Chia SJ, Tan EK, Chao YX. Historical Perspective: Models of Parkinson's Disease. Int J Mol Sci. 2020;21:2464.

11. Miller N, Greene K, Dydinski A, Gerlai R. Effects of nicotine and alcohol on zebrafish (Danio rerio) shoaling. Behav Brain Res. 2013;240:192-6.

12. Pippal JB, Cheung CM, Yao YZ, Brennan FE, Fuller PJ. Characterization of the zebrafish (Danio rerio) mineralocorticoid receptor. Mol Cell Endocrinol. 2011;332:58-66.

13. Steenbergen PJ, Richardson MK, Champagne DL. The use of the zebrafish model in stress research. Prog Neuropsychopharmacol Biol Psychiatry. 2011;35:1432-51.

14. Caroline H Brennan. Zebrafish behavioural assays of translational relevance for the study of psychiatric disease. Rev Neurosci. 2011;22:37-48.

15. Bianchi, Cooper, Gordon, Heemskerk, Hodes, Koob, Koroshetz, Shurtleff, Sieving, Volkow, Churchill, Ramos. Neuroethics for the National Institutes of Health BRAIN Initiative. J Neurosci. 2018;38:10583-10585.

16. Loeb JM, Hendee WR, Smith SJ, Schwartz MR. 1989. Human vs animal rights. In defense of animal research. JAMA 262:2716–2720.

17. Encyclopedia Britannica. 2013. Claude Bernard. Available at http://www. britannica.com/ EBchecked/topic/62382/Claude-Bernard

18. Carlos Zaragoza, Carmen Gomez-Guerrero, Jose Luis Martin-Ventura, Luis Blanco-Colio, Begoña Lavin, Beñat Mallavia, Carlos Tarin, Sebastian Mas, Alberto Ortiz, and Jesus Egido. Animal Models of Cardiovascular Diseases. J Biomed Biotechnol. 2011; 2011: 497841.

19. Gerald Brandacher, Johanna Grahammer, Robert Sucher, Wei-Ping Andrew Lee. Animal models for basic and translational research in reconstructive transplantation. Birth Defects Res C Embryo Today. 2012;96:39-50.

20. Ramin Sedaghat Herati and E. John Wherry. What Is the Predictive Value of Animal Models for Vaccine Efficacy in Humans? Cold Spring Harb Perspect Biol. 2018 Apr; 10(4): a031583.

21. B.J. Wilkesa and M.H. Lewis. The Neural Circuitry of Restricted Repetitive Behavior: Magnetic Resonance Imaging in Neurodevelopmental Disorders and Animal Models. Neurosci Biobehav Rev. 2018; 92: 152–171.

22. Brittanie Partridge and John H. Rossmeisl, Jr. Companion animal models of neurological disease. J Neurosci Methods. 2020; 331: 108484.

23. Russell, W.; Burch, R. The Principles of Humane Experimental Technique; Methuen& Co. Ltd.: London, UK, 1959.

24. Palmer A, Message R, Greenhough B. Edge cases in animal research law: Constituting the regulatory borderlands of the UK's Animals (Scientific Procedures) Act. Stud Hist Philos Sci. 2021;90:122-130.

25. Kevin A Ford. Refinement, Reduction, and Replacement of Animal Toxicity Tests by Computational Methods. ILAR J. 2016;57:226-233.

26. Richard M. A. Parker, William J. Browne. The Place of Experimental Design and Statistics in the 3Rs. ILAR Journal, Volume 55, Issue 3, 2014, 477–485.

27. Ian J.H. Duncan. The changing concept of animal sentience. Applied Animal Behaviour Science. Volume 100, Issues 1–2, 2006, 11-19.

28. Franz P Gruber, Thomas Hartung. Alternatives to animal experimentation in basic research. ALTEX. 2004;21 Suppl 1:3-31.

29. Marijana Vučinić, Saša Trailović, Zoran Todorović and Milica Prostran. Ethics of animal use in preclinical phase of drug testing. Bioethics and Pharmacology: Ethics in Preclinical and Clinical Drug Development, 2012: 15-31.

30. National Academies of Sciences, Engineering, and Medicine. 1991. Science, Medicine, and Animals. Washington, DC: The National Academies Press.

31. Hemmer B, Archelos JJ, Hartung HP. New concepts in the immunopathogenesis of multiple sclerosis. Nat Rev Neurosci 2002;3:291–301.

32. Noseworthy JH, Lucchinetti C, Rodriguez M, Weinshenker BG. Muliple sclerosis. New Engl J Med 2000;343:938 –946.

33. Zamvil S, Steinman LS. EAE and autoimmunity. Annu Rev Immunol 1990;8:579– 621.

34. Owens T, Sriram S. The immunology of multiple sclerosis and its animal model, experimental allergic encephalomyelitis. Neurol Clin 1995;13:51–73.

35. Rivers TM, Schwentker FF. Encephalomyelitis accompanied by myelin destruction experimentally produced in monkeys. J Exp Med 1935;61:689 –701.

36. Steinman L. Assessment of animal models for MS and demyelinating disease in the design of rational therapy. Neuron 1999;24:511–514.

37. Martin R, McFarland H. Experimental immunotherapies for multiple sclerosis. Semin Immunopathol 1996;18:1–24.

38. Jyothi MD, Flavell RA, Geiger TL. Targeting autoantigenspecific T cells and suppression of autoimmune encephalomyelitis with receptor-modified T lymphocytes. Nat Biotechnol 2002;20:1215–1220.

39. Sospedra M, Martin R. 2005. Immunology of multiple sclerosis. Annu Rev Immunol 23: 683–747.

40. Lassmann H, Bruck W, Lucchinetti CF. 2007. The immunopathology of multiple sclerosis: An overview. Brain Pathol 17: 210–218.

41. Carolyn Goldschmidt and Marisa P. McGinley. Advances in the Treatment of Multiple Sclerosis. Neurol Clin. 2021;39: 21–33.

42. Steinman, L. and Zamvil, S.S. (2005) Virtues and pitfalls of EAE for the development of therapies for multiple sclerosis. Trends Immunol. 26, 565–571.

43. Phillips, K.A., Bales, K.L., Capitanio, J.P., Conley, A., Czoty, P.W., ′t Hart, B.A., Hopkins, W., Hopkins, W.D., Hu, S.L., et al. Why primate models matter. Am. J. Primatol. 2014, 76, 801–827.

44. Maestripieri, D. The Past, Present, and Future of Primate Psychology. In Primate Psychology; Maestripieri, D., Ed.; Harvard University Press: Cambridge, MA, USA, 2003; pp. 1–16.

45. European Commission. Seventh Report on the Statistics on the Number of Animals used for Experimental and other Scientific Purposes in the Member States of the European Union. Available online http://eur-lex. europa.eu/legal-content/EN/TXT/?uri=CELEX:52013DC0859

46. Lankau, E.W.; Turner, P.V.; Mullan, R.J.; Galland, G.G. Use of nonhuman primates in research in North America. J. Am. Assoc. Lab. Anim. Sci. 2014, 53, 278–282.

47. Zhang, X.L.; Pang, W.; Hu, X.T.; Li, J.L.; Yao, Y.G.; Zheng, Y.T. Experimental primates and non-human primate (NHP) models of human diseases in China: Current status and progress. Zool. Res. 2014, 35, 447–464.

48. Goodall, J. Tool-using and aimed throwing in a community of free-living chimpanzees. Nature 1964, 201,1264–1266.

49. Liu, Q; Fragaszy, D.M.; Visalberghi, E. Wild capuchin monkeys spontaneously adjust actions when using hammer stones of different mass to crack nuts of different resistance. Am. J. Phys. Anthrop. 2016, 161, 53–61.

50. Kawamura, S. The process of sub-culture propagation among Japanese macaques. Primates 1959, 2, 43–60.

51. Eshchar, Y.; Izar, P.; Visalberghi, E.; Resende, B.; Fragaszy, D. When and where to practice: Social influences on the development of nut-cracking in bearded capuchins (Sapajuslibidinosus). Anim. Cogn. 2016, 19: 605–618.

52. Goodall, J. Tool-using and aimed throwing in a community of free-living chimpanzees. Nature 1964, 201:1264–1266.

53. Liu, Q.; Fragaszy, D.M.; Visalberghi, E. Wild capuchin monkeys spontaneously adjust actions when using hammer stones of different mass to crack nuts of different resistance. Am. J. Phys. Anthrop. 2016:161, 53–61.

54. Freeman, H.D.; Gosling, S.D. Personality in nonhuman primates: A review and evaluation of past research. Am. J. Primatol. 2010, 72, 653–671.

55. Cheney, D.; Seyfarth, R.; Smuts, B. Social relationships and social cognition in nonhuman primates. Science 1986, 234, 1361–1366.

56. Smuts, B.B.; Cheney, D.L.; Seyfarth, R.M.; Wrangham, R.W. Primate Societies; University of Chicago Press: Chicago, IL, USA, 2008.

57. Sterck, E.H.; Goossens, B.M. The meaning of "macaque" facial expressions. Proc. Natl. Acad. Sci. USA 2008,105, E71.

58. Yang, B.; Anderson, J.R.; Li, B.G. Tending a dying adult in a wild multi-level primate society. Curr. Biol. 2016, 26, 403–404.

59. Kathleen M Conlee, Andrew N Rowan. The case for phasing out experiments on primates. Hastings Cent Rep.2012;Suppl:S31-4.

60. European Commission. Special Eurobarometer: Science and Technology Report. Available online:http://ec.europa.eu/commfrontoffice/publicopinion/archives/ebs/ebs_340_en.pdf.

61. Beauchamp, T.L.; Childress, J.F. Principles of Biomedical Ethics, 4th ed.; Oxford University Press: New York, NY, USA, 2005.

62. Nardini, C. The ethics of clinical trials. Ecancermedicalscience 2014, 8, 387.

63. Gluck, J.P. Moving beyond the welfare standard of psychological well-being for nonhuman primates: The case of chimpanzees. Theor. Med. Bioeth. 2014, 35, 105–116.

64. Thomas, D. Laboratory animals and the art of empathy. J. Med. Ethics 2005, 31, 197–202.

65. Kant, I. Critique of Practical Reason, 2nd ed.; Cambridge University Press: Cambridge, UK, 2015.; nt, I. Groundwork of the Metaphysics of Moral; Cambridge University Press: Cambridge, UK, 2012.

66. Ross, W.D. The Right and the Good; Clarendon Press: Oxford, UK, 1930.

67. Mandal, J.; Ponnambath, D.K.; Parija, S.C. Utilitarian and deontological ethics in medicine. Trop. Parasitol. 2016, 6, 5–7.

68. Donaldson, T.M. Ethical resources for the clinician: Principles, values and other theories. In Contemporary Issues in Bioethics; Beauchamp, T.L., Walters, L., Kahn, J.P., Mastroianni, A.C., Eds.; Borden Institute: Washington, DC, USA, 2003; pp. 15–38.

69. Kant I. 2002. Groundwork for the metaphysics of morals. New Haven (CT): Yale University Press.

70. Tannenbaum J. 2000. Animal rights and animal research. In Kraus AL, Renquist D (eds). Bioethics and the use of laboratory animals. Dubuque (IA): Gregory C Benoit Publishing.

71. Wolfensohn S, Lloyd M. 2013. Handbook of laboratory animal management and welfare, 4th ed, p 37–38. Oxford (UK): Wiley and Blackwell.

72. Bentham, J., An Introduction to the Principles of Morals and Legislation; The Clarendon Press: Oxford, UK, 1823.

73. Mooney, S. The Moral and Practical Considerations of the Use of Antibiotics in Concentrated Animal Feeding Operations on Non-Human and Human Populations. 2014. Available online:https://scholarsbank.uoregon.edu/xmlui/bitstream/handle/1794/18275/Thesis%20

74. Cong L, Ran FA, Cox D, Lin S, Barretto R, Habib N, Hsu PD, Wu X, Jiang W, Marraffini LA, Zhang F. Multiplex genome engineering using CRISPR/Cas systems. Science. 2013;339(6121):819-23.

75. Jennings CG, Landman R, Zhou Y, Sharma J, Hyman J, Movshon JA, Qiu Z, Roberts AC, Roe AW, Wang X, Zhou H, Wang L, Zhang F, Desimone R, Feng G, Opportunities and challenges in modeling human brain disorders in transgenic primates. Nat Neurosci. 2016 26;19:1123-30.

76. DeGrazia D. Nonhuman Primates, Human Need, and Ethical Constraints. Hastings Cent Rep. 2016 Jul;46(4):27-8.

ETHICAL ASPECTS OF THE USE OF GENETICALLY MODIFIED FEED IN LIVESTOCK

Ivica Kelam

Faculty of Education/Faculty of Kinesiology,
Josip Juraj Strossmayer University of Osijek, Croatia

Abstract

In 2019, 190.4 million hectares of biotech crops were planted by up to 17 million farmers in 29 countries. Over 70% of harvested genetically modified biomass is fed to food-producing animals, making them the primary consumers of genetically modified crops for the past 20 years. However, genetically modified crops serve primarily as feed for livestock and put tremendous pressure on the environment. In this paper, we will analyse the following issues. First, how the transformation of livestock driven by genetically modified feeds leads to unsustainable practices that negatively affect climate change, lead to environmental degradation and harm the welfare of farm animals. Second, we analyse the significant increase in the use of glyphosate in the sowing of genetically modified crops and the increase in the residues of glyphosate in the human and livestock food chain. Third, we will give a brief overview of animal feeding trials studies through the impact of glyphosate on animal health and welfare. In conclusion, we will highlight the ethical questionability of using the genetically modified feed in livestock.

Keywords: genetically modified crops, glyphosate, livestock, ethics, feed

Introduction

Since the discovery of genetic modification of plants in the early 1980s, controversy has begun that is still unfinished. Critics have expressed doubts about the justification of genetic modification, pointing out that it will only further increase the power of biotech corporations to farmers. However, genetically modified (GM) plant seeds for food and feed production continuously increases worldwide. The latest data indicated that GM crops were grown on 191.7 million hectares worldwide in 2018. In 2018, GM soybean occupied 50% of the global area under modified crops [1]. GM soybeans have remained the main such crop since 1996 when the first commercialised genetically modified crop seeds came to notice. Throughout the 23 years since, soybeans have held the top position regarding the area covered by GM crop production [2]. Over time, it has become apparent that most genetically modified crops, primarily GM soy and GM corn, are used as livestock feed. The use of genetically modified crops, primarily soybeans and, to a lesser extent, corn as feed, entails whether it is ethically permissible to feed animals with genetically modified soybeans. In the paper, we will try to answer this dilemma. In the first part of the paper, we will deal with the transformation of livestock farming, which has transformed from small farms to factory farms owned by corporations in the last few decades. Genetically modified feeds drove the transformation of the livestock sector, resulting in unsustainable practices that negatively affect climate change, lead to environmental degradation and harm the welfare of farm animals. In the second part, we analyse the significant increase in the use of glyphosate in the sowing of genetically modified crops and the increase in glyphosate residues in the human and livestock food chain since the use of glyphosate is key to the commercial success of genetically modified crops. Glyphosate residues have environmental and health costs, and we will write about them in this chapter. In the third part of the paper, we will give a brief overview of animal feeding trials studies through the impact of glyphosate on animal health and welfare because there are many studies. We will concentrate on the latest studies, especially those indicating possible harmful effects on animal health and welfare. In conclusion, based on the above, we will consider the ethics of using the genetically modified feed in livestock.

The ethical aspects of livestock farming transformation

In considering the ethical aspects of the use of genetically modified feed in animal nutrition, it is necessary to look at the significant transformation of animal husbandry in recent decades, which was accompanied by a significant increase in consumption of animal products, primarily through the consumption of meat, eggs and dairy products. As a result, there has been the collapse of small family-owned farms and the rapid consolidation of large corporate industrial farms that occupy vast estates and vast numbers of animals in overcrowded and unhygienic conditions. Small family farms, which were the primary source of meat and other animal products for customers, were far more suitable for more sustainable and environmentally friendly practices and cared for animal welfare. However, factory-raised animals combined with state subsidies have enabled cheaper production and lower prices for customers, with which many small local producers cannot compete. Furthermore, factory farms are designed to maximise profits, which has made livestock far more efficient by introducing new technologies and mechanical innovations, as noted in the MacDonald and McBride report: „Bigger and faster equipment allows one producer to cultivate, sow, fertilise, spray or harvest more hectares; house and feed more cattle or poultry; or milk more cows in one day" [3, 8]. Unfortunately, the price of these innovations is paid by the welfare of animals and farmworkers' health, according to numerous scientific reports [4]. The business model and success of the modern factory farm relies on genetically modified feed for its animals and various chemicals and pharmaceuticals to mitigate the spread of disease and facilitate the growth of its animals to market size in the shortest possible time. We emphasise that today's factory farming would be impossible without sufficiently cheap GM feed and extensive use of antibiotics to prevent disease in animals in overcrowded pens.

In conclusion, factory farming is highly unethical because of the inhumane treatment of animals and is destructive to the environment by producing vast amounts of manure [5], contributing to climate change through the problem of CO_2 and methane emissions [6]. A particular problem is the massive water consumption since, according to UNESCO calculations, 15,000 litres of water is required to produce 1 kg of beef [7]. Additionally encourages deforestation of forests due to the planting of genetically modified soybeans, emphasising the problem of deforestation of the Amazon rainforest [8]. GM soy is the second-largest agricultural driver of deforestation after cattle products. According to a 2013 study for the European Commission, soy expansion was responsible for nearly half of the deforestation embodied in products imported into the European Union between 1990 and 2008 [9]. Furthermore, finally, the growing problem that glyphosate poses to human and livestock health we will analyse in the next chapter.

The rise and role of glyphosate in the human and livestock food chain

In considering the ethical aspects of genetically modified food and feed, it is necessary to include the crucial role of glyphosate and other herbicides in producing genetically modified crops. Without glyphosate, there are no genetically modified crops in this form. This chapter will pay special attention to the role of glyphosate-tolerant (GT) soy, as it dominates the world soybean market. Furthermore, we will emphasise the importance and possible dangers arising from the dependence of the livestock sector on genetically modified soy mainly used as feed. Glyphosate tolerant (GT) soy is a globally dominant genetically modified (GM) plant and trait combination. About 77% of the global soybean production comes from GT soybean. The leading soybean-producing countries are Argentina, Brazil and the United States, with adoption rates ranging from 94% to 100% of biotech crops, mainly glyphosate-tolerant varieties. [10]. According to Brokes and Barefoot, the implementation of GT soybean in the agricultural production of these countries has contributed to increasing the gross income of agricultural holdings mainly by reducing production costs [11]. In short, GT soybean agriculture consists of patented genetically modified soybean seeds and herbicides (most commonly glyphosate) - which allows farmers to destroy weeds by spraying herbicides during the growing season. Soybean plants are genetically modified to tolerate herbicides. In his paper, Benbrook links the extremely rapid growth in sales and use of glyphosate-based herbicides (GBH) such as Roundup with the successful sale of glyphosate-resistant soybean seeds [12]. Thus, according to official USDA data, 349 million metric tons (MT) of soybeans in the world were produced in the 2016-2017 season, of which GT varieties contributed about 270 million MT [13]. These figures are not surprising since soy is the dominant ingredient in animal feed globally. Soy is the largest source of plant protein in animal feed, potentially affecting many plant material and meat-eaters consumers. This problem is enormous in the EU, as the EU imports 95% of its soya beans and soya meal, or 36.1 million tons of soybean equivalent per year for livestock feed [14]. Over 95% of these imports come from five countries in the Americas, where GM technology adoption is between 94% and 100%. The biggest soya exporters to the EU are the USA, Brazil, and Argentina - the very same three countries which are also the leading food providers to the EU and the leading GM adopters in general. Consequently, GM soy which feeds livestock in the EU, ultimately ends up on the plates of European consumers. According to a WWF study, in Europe, the average citizen unconsciously consumes 57 kg of soy per year (93 per cent of total consumption) through animal foods. According to a study, the soy content per kilogram of livestock product (in retail weight) is highest for broilers (1,089 grams per kg), farmed fish (738 grams per kg), pork (508 grams per kg) and beef (456 grams per kg). An average EU-28-citizen roughly consumes 53 kg of pork, poultry and beef (in retail weight), 214 eggs, 134 kg of dairy products and 3 kg of farmed fish (in retail weight) per year. This corresponds to a daily intake of about 146 grams of meat, half an egg, 367 grams of dairy products and 8 grams of farmed fish [15].

Because of all the above, GT soy should be subject to intense scrutiny from leading agricultural and food safety authorities, such as the European Food Safety Authority (EFSA), the US Food and Drug Administration (FDA), the World Health Organization (WHO) and especially from independent researchers. Despite significant amounts of published scientific research on the safety of GM plants, they are still controversial to a large part of the public, especially in Europe. However, a majority of the safety studies of GM plants have so far concluded that those plants can be as safe and nutritious as conventional plants [16]. Most studies on the safety of GM products have been conducted and financed by biotechnology companies such as Monsanto, which are responsible for the commercial introduction of these new varieties on the market, and therefore cause public distrust in the results of these studies [17]. Diels et al. analysed 94 articles on the presumed health risks of GM products. The results showed that individual papers' choice of research results was significantly influenced by financial and/or professional conflicts of interest [18].

Moreover, these results are confirmed by cases in which some independent researchers have been subjected to personal attacks and threats of legal action by owners of patent rights to GM seeds, i.e. biotech corporations, trying to prevent the publication of critical results on GM crops [19, 20, 21]. However, glyphosate (commonly known as "Roundup") was placed on the market as early as 1974. Since the late 1970s, the amount of glyphosate-based herbicides (GBH) applied has increased approximately 100-fold, with particularly explosive growth following the sowing of genetically modified crops resistant to glyphosate since 1994 [22]. Moreover, according to Benbrook, glyphosate became the most widely used herbicide globally by 2015, with global applications increasing 15-fold from 1990 to 2014 [12]. Osteen and Fernandez-Cornejo, in their paper, analyse rapid glyphosate increase in use on four major crops in the US – corn, cotton, soy, and wheat. According to them, glyphosate accounted for only 1% of herbicides sprayed on those crops in 1982, slowly climbing to 4% in 1995, and after the introduction of GM crops shot up to 33% in 2005 and 40% in 2012 [23]. As a result, glyphosate has become widely present in the global soybean supply, and the pragmatic regulatory response has been to adjust these changes in glyphosate use by significantly increasing daily intake levels that are considered tolerant [24]. Although GT soybeans are not grown in Europe at all, their widespread cultivation in the US, Argentina and Brazil as leading exporters of GT soybean feed to the EU has resulted in significant glyphosate residues in soybean products imported into Europe for food and feed [25]. Therefore, the following question arises: how much glyphosate enters the food chain of humans and livestock through the glyphosate residue? Bøhn et al. in their paper assume that the glyphosate residue levels found in GT soybeans from the USA were representative of what is found on the international market, i.e., an average of ~9.0 mg/kg, the amount of glyphosate entering the food chain would be 9 g/tonne of soy. Their calculations would add up to 2430 tonnes of glyphosate residues from the 270 million MT of GT soy produced globally in the 2016/2017 season [26]. Data on glyphosate residue from Brazil and Argentina are significantly higher, suggesting that these estimates from the USA are rather conservative. For example, Brazilian samples showed average glyphosate residue levels at 38.5 mg/kg [27], i.e. almost twice the maximum accepted residue levels (MRLs) as stated in Codex and the EU [28]. While in Argentina, average and maximum residue levels were measured at 31.7 mg/kg and 72.8 mg/kg, respectively [29].

Furthermore, comparing this data with data from corporate field trials, which show far lower levels of glyphosate residues, and only about 1 mg/kg, i.e. several tens times less than the concentrations from the above countries [30]. On the example of the discrepancy of these glyphosate residue data between actual levels in the USA, Brazil and Argentina fields and corporate field trials, it is not surprising that the public is still sceptical about the safety and health of GM crops. It is unacceptable that the results and data of corporate field trials, which regulatory agencies use to assess GT soybean varieties' safety, are sprayed according to agricultural practices that differ significantly from modern commercial agriculture.

Consequently, these corporate field trials produce unrealistic and irrelevant patterns, leading regulatory agencies to underestimate potential risks due to insufficient data systematically. For example, Cuhra, in his paper, points out that most feeding studies used unsprayed test plants to assess the quality of GT soybeans (and other GT plants) and concluded: „Despite decades of risk assessments and research in this field, specific unanswered questions relating to safety and quality aspects of food and feed from GM crops need to be addressed by regulators. Independent research provides important supplementary insight" [32, 1]. Unfortunately, GT soybean varieties currently imported into the EU have been approved based on these insufficient corporate studies and their data. A good example of this unethical practice is a 90-day rat feeding study to test the quality of triple-resistant GT soybeans, commercially named 'Enlist' (which is resistant to herbicides based on glyphosate, glufosinate-ammonium and 2,4-D). The scientists found that this GT soybean variety was sprayed with lower doses of glyphosate and 2.4-D than recommended for commercial use in the fields and was not sprayed with ammonium glufosinate [32]. Although it has been scientifically established that glufosinate-ammonium can cause central nervous system damage, including memory loss in mice [33], EFSA is therefore considered a high-risk chemical for humans [34]. Nevertheless, despite all the shortcomings of these tests, EFSA has accepted corporate data and recently approved this GT soybean for import into the EU [35].

In addition, a particular cause for concern presents adjuvants (additives and wetting agents) added to herbicides used in biotech agriculture.

According to some studies, these adjuvants can increase toxicity by a factor of up to 1000 times compared to a single active substance [36]. For this reason, the use of particularly problematic aids, such as polyethoxylated tallowamine (POEA), is restricted or banned in several EU countries. For example, POEA is a significant component of Glyphosate based herbicide (GBH) toxicity, especially for aquatic organisms such as protozoa, crustaceans, frogs, toads, and fish [37]. However, the major problem is that the use of POEA is not banned in leading countries where GT plants are grown, such as Argentina, where herbicides applied to GT soybean fields typically contain about 15% POEA [38].

Consequently, GT soybeans imported into the EU from Argentina are sprayed with formulations not allowed in the EU. Although glyphosate is now approved for use in the EU by the end of 2022, a detailed examination of residues from imported food and animal products is necessary if the plants are grown under a different regulatory regime / agricultural practice [39]. Because according to the EFSA Pesticides Panel, existing data are insufficient to assess the health hazards of GT plants in the consumption phase, which is an additional cause for concern [40].

Review of animal feeding trials studies through the impact of glyphosate on animal health and welfare

The remarkable growth in the use of glyphosate, which has been closely linked to the sowing of genetically modified crops over the past 25 years, has raised concerns about its impact on human health, mainly its association with various tumours, most notably non-Hodgkin's lymphoma. In light of these concerns, a 2015 study by the World Health Organization's International Agency for Research on Cancer (IARC) finally declared glyphosate 'likely carcinogenic' because it found "limited" evidence of cancer links in studies of human exposures, mostly agricultural-related, that had been published since 2001. However, IARC said studies in laboratory animals showed "sufficient" evidence that glyphosate can cause cancer. Research also showed "strong" evidence that glyphosate caused DNA and chromosomal damage in human cells [41]. A new classification of glyphosate by the WHO has sparked a heated debate on the toxicological profile of glyphosate, as the biotechnology industry has vigorously challenged the WHO decision. However, Krimsky and Gillam point out corruption in science in their paper because: "The court-released discovery documents obtained from litigation against Monsanto over its herbicide Roundup and through Freedom of Information Act requests (requests to regulatory agencies and public universities in the United States). We sought evidence of corporate malfeasance and undisclosed conflicts of interest with respect to issues of scientific integrity. The findings include evidence of ghostwriting, interference in journal publication, and undue influence of a federal regulatory agency"[42, 318]. The Krimsky and Gillam papers prove that the biotechnology industry, led by Monsanto, influenced the results of studies to prove the safety of glyphosate.

In the following, we will look at studies that indicate possible harmful health effects of glyphosate on animal and human health. In Denmark, a team of scientists from the University of Aarhus, led by scientist Martin Tang Sørensen, conducted an extensive research project from January 2017 to 2020. The project will examine whether glyphosate residues originating from herbicides used in feed production (for instance, Roundup) can affect the health and productivity of farm animals. Research has shown that Roundup can affect the gastrointestinal health of mammals. In particular, they found that when sprayed on genetically modified soybean food, glyphosate came into direct contact with livestock through the diet and endangered their health. Besides that, they found that the pesticide destroyed gastrointestinal tract bacteria that acted as guarantors of the natural balance of the intestines of animals. [43]

In conclusion, Sørensen et al. point out the lack of scientific studies as one of the most significant issues stating: „Literature, addressing potential consequences of the glyphosate chelating property for livestock health and productivity, basically do not exist" [43, 7]. In addition to the problems that glyphosate can cause with livestock, recent research suggests that glyphosate can also be harmful to fish. The conclusions of the study of the impact of glyphosate on tilapia fish from 2021 stated: "The present study demonstrated an underlying toxicity mechanism of Gly long-term exposure in tilapia. Long-term exposure to Gly decreased antioxidative ability, disturbed hepatic metabolism, induced inflammation and suppressed immune response. It was worth noting that Gly exposure impaired Nrf2 pathway and downregulated its downstream genes, which further reduced antioxidative ability. Meanwhile, Gly exposure activated NF-κB pathway, and promoted the production of pro-inflammatory factors, such as TNF-α and IL-1β, which worsened inflammation. These data may further enrich the toxicity mechanism of Gly in fish, which may provide evidence for the risk assessment of Gly in aquatic environment" [44, 8]. In their review of feeding animal studies, Matovu and Alçiçek point out: „This review aimed to evaluate the probable transfer and accumulation of tDNA/proteins from transgenic feeds in animal samples (ruminant and non-ruminant) by evaluating the available experimental studies published scientifically. This study found that the tDNA/protein is not completely degraded during feed processing and digestion in Gastro-Intestinal Tract (GIT). In large ruminants (cattle), tDNA fragments/proteins were detected in GIT digesta, rumen fluid, and faeces. In small ruminants (goats), traces of tDNA/proteins were detected in GIT digesta, blood, milk, liver, kidney, heart and muscle. In pigs, they were detected in blood, spleen, liver, kidney, and GIT digesta. In poultry, traces were detected in blood, liver and GIT digesta but not in meat and eggs. Notwithstanding some studies that have shown transfer of tDNA/ protein fragments in animal samples, we cannot rely on these few studies to give general evidence for transfer into tissues/fluids and organs of farm animals. However, this study clearly shows that transfer is possible. Therefore, intensive and authentic research should be conducted on GM plants before they are approved for commercial use, investigating issues such as the fate of tDNA or proteins and the effects of feeding GM feed to livestock" [45, 1].

When we talk about the need and usefulness of animal feeding studies in authorising GM crops, we need to be aware that there is no international consensus on the necessity for animal feeding studies. Thus, for example, the study of Bartholomaeus et al. [46] published in the prestigious journal Critical Reviews in Toxicology challenges such studies' value, especially as a permanent requirement. Bartholomaeus et al. have summarised their opposition as follows: "(1) commercial production of GM (GE) crops can and does produce unintended, unexpected, and unpredictable compositional effects unrelated to both the parent line of the crop and to the inserted transgene; (2) the unintended, unexpected, compositional effects resulting from the production of GE crop varieties are potentially of toxicological significance; (3) compositional analysis of new GE varieties is insufficiently sensitive to reliably detect these differences at levels of toxicological significance; (4) toxicity studies on whole feed are capable of detecting toxicologically significant differences that would be missed by agronomic and compositional analysis" [46, 4]. Bartholomaeus et al. reviewing various data from feeding trials conclude that animal feeding trials "in fact they did not add value, were unnecessary, and scientifically unjustifiable". [46, 14] Finally, we mention a similar study by Kuiper et al. [47] in which they strongly advocate against any animal feeding study (except on a case-by-case basis), stating: "Routine testing should not be required since, due to apparent weaknesses in the approach, it does not add to current risk assessment of GM foods. Moreover, the demand for routine testing using animals is in conflict with the European Union (EU) Commission's efforts to reduce animal experimentation. Regulating agencies in the EU are invited to respect the sound scientific principles applied to the risk assessment of foods derived from GM plants and not to interfere in the risk assessment by introducing extra requirements based on pseudo-scientific or political considerations" [47, 781]. Based on all of the above, we must ask the logical question, did animal nutrition studies have any value in assessing the safety of GM crops? Scientists Herman and Ekmay offer their solution and suggest: „While the justification of animal feeding studies to detect unexpected effects may be inadequately supported, there may be better justification to conduct such studies in specific cases to investigate the consequences of expected compositional effects including expression of transgenic proteins. Such studies may be justified when (1) safety cannot reasonably be predicted from other evidence, (2) reasonable hypothesis for adverse effects are postulated, (3) the compositional component in question cannot be isolated or enriched in an active form for inclusion in animal feeding studies, and (4) reasonable multiples of exposure can be accomplished relative to human diets. The study design for whole-food animal-feeding studies should be hypotheses-driven, and the types of data collected should be consistent with adverse effects that are known to occur from dietary components of biological origin" [48, 171]. This whole situation around the role of glyphosate residues, animal feeding trials, and the impact of glyphosate residues on human and animal health Torretta et al. in their paper summarises: "Despite numerous studies regarding the dangers resulting from the extensive use of glyphosate, it is not possible to attribute a clear and unambiguous definition to glyphosate, especially regarding its potentially harmful effects on humans. Resuming and analysing each treated subject, it is concluded that the massive use of glyphosate, given its chemical-physical properties and its presence in many commercial products, even in the form of more toxic mixtures than the single molecule, should be reduced. This could be achieved by implementing resources and funding dedicated to alternative solutions to the herbicide. However, diffusion is unfortunately not yet advanced due to the huge economic interest in the market for herbicides. It is also evident that new studies and independent research must be performed in order to clearly define the seriousness of glyphosate exposure to carcinogenicity and genotoxicity. In fact, there is a too great a discrepancy between the opinions of the various scientific institutions, mainly because of their different economic and social interests. The controversy and the debate are expected to continue. It is the duty of the institutions to place the "precautionary principle" before economic interests, namely the protection of citizens and the environment from exposure to a substance whose side effects are not yet known." [49, 16].

Conclusion

We cannot give a simple answer to whether it is ethically permissible to use the genetically modified feed in livestock. Nor can we say that genetically modified feed is not ethically permissible just because it is a matter of genetic modification technology as if genetic modification technology is unethical in itself. In assessing the ethics of genetically modified feed, we need to look at this problem multidimensionally, considering health aspects and economic, social, environmental, and many others. That is why in our paper, we first looked at the transformation of livestock farming and the role of genetically modified feed in it. However, the transformation of livestock farming began before the sowing of genetically modified crops. Global access to plant proteins from GT soybeans has enabled further corporate takeovers of the livestock sector, factory farms' emergence, and all harmful side effects, from manure disposal,

CO2 and methane emissions to deforestation to plant new GT soybean plantations. Therefore, we can conclude that today's livestock factory farming is ethically questionable, as it is environmentally unsustainable and causes immeasurable suffering to animals, and genetically modified feed further deepens these problems. In addition, the use of glyphosate in the cultivation of genetically modified crops is a major environmental, health and ethical issue. The ubiquity of glyphosate results in a significant increase in glyphosate residues in the environment and bioaccumulates in the tissues of animals and humans. Bioaccumulation of glyphosate residues in livestock poses a potentially significant health problem, as GT soy is a significant source of protein for a large portion of livestock.

Consequently, the question arises is there a potential danger to human health when consuming products of animal origin, meat, eggs and dairy products? Moreover, is it ethically permissible to feed livestock with the genetically modified feed? Unfortunately, we do not have a definitive answer to these questions, as long-term scientific studies on the impact of glyphosate residues on livestock health have not been conducted. Therefore, we conclude that the use of genetically modified feed in livestock is ethically questionable, as not all effects of glyphosate residue on animal health and welfare are yet known.

References

1. International Service for the Acquisition of Agri-biotech Applications (ISAAA). Global Status of Commercialized Biotech/GM Crops in 2018: Biotech Crops Continue to Address the Challenges of Increased Population and Climate Change. Available online: https://www.isaaa.org/resources/publications/briefs/54/download/isaaa-brief-54-2018.pdf.
2. Sieradzki, Z., Mazur, M., Król, B., & Kwiatek, K. Prevalence of genetically modified soybean in animal feedingstuffs in Poland. Journal of Veterinary Research, 2021, 65 (1), 93.
3. MacDonald, James M., and William D. McBride. The Transformation ofU.S. Livestock Agriculture: Scale, Efficiency, and Risks. A Report from the Economic Research Service. No. 43. United States Department of Agriculture, 2009. Available online: https://www.ers.usda.gov/webdocs/publications/44292/10992_eib43.pdf?v=0.
4. Blanchette, A. Living Waste and the Labor of Toxic Health on American Factory Farms. Medical Anthropology Quarterly (2019). doi:10.1111/maq.12491.

5. Hribar, C. Understanding Concentrated Animal Feeding Operations and Their Impact on Communities. Ed. Mark Schultz. Ohio: The National Association of Local Boards of Health (2010). Available online: https://www.cdc.gov/nceh/ehs/docs/understanding_cafos_nalboh.pdf.

6. Tauseef, S.M., M. Premalatha, Tasneem Abbasi, and S.A. Abbasi. Methane Capture from Livestock Manure. Journal of Environmental Management, 2013, 117 (15); 187–207. DOI: 10.1016/j.jenvman.2012.12.022

7. UNESCO. 'Good water, water to "eat". What is virtual water?' http://www.unesco.org/new/fileadmin/MULTIMEDIA/FIELD/Venice/pdf/special_events/bozza_scheda_DOW04_1.0.pdf. Available online: http://www.unesco.org/new/fileadmin/MULTIMEDIA/FIELD/Venice/pdf/special_events/bozza_scheda_DOW04_1.0.pdf.

8. Richards, P. D., Myers, R. J., Swinton, S. M., & Walker, R. T. Exchange rates, soybean supply response, and deforestation in South America. Global Environmental Change, 2012, 22(2), 454–462. doi:10.1016/j.gloenvcha.2012.01.0

9. European Commission. The impact of EU consumption on deforestation: Comprehensive analysis of the impact of EU consumption on deforestation. Available online: https://ec.europa.eu/environment/forests/pdf/1.%20Report%20analysis%20of%20impact.pdf

10. James, C. Global Status of Commercialized Biotech/GM Crops: 2017; ISAAA: Ithaca, 2017.

11. Brookes, G.; Barfoot, P. Farm income and production impacts of using GM crop technology 1996–2015. Gm Crop. Food 2017, 8, 156–193.

12. Benbrook, C.M. Trends in glyphosate herbicide use in the United States and globally. Environ Sci Eur, 28, 3 (2016) https://doi.org/10.1186/s12302-016-0070-0

13. USDA. World Agricultural Supply and Demand Estimates; United States Department of Agriculture: Washington, 2018; pp. 1–40.

14. EU market observatory, Crops market observatory. Available online: https://circabc.europa.eu/sd/a/2c8378ab-c686-449d-9dd1-65371ab30889/Oilseeds-dashboard_en.pdf.

15. Kroes, H.; Kuepper, B.; Mapping the soy supply chain in Europe. Available online: https://wwfint.awsassets.panda.org/downloads/mapping_soy_supply_chain_europe_wwf_2015.pdf.

16. Domingo, J.L.; Bordonaba, J.G. A literature review on the safety assessment of genetically modified plants. Environ. Int. 2011, 37, 734–742.

17. Domingo, J.L. Safety assessment of GM plants: An updated review of the scientific literature. Food Chem. Toxicol. 2016, 95, 12–18.

18. Diels, J.; Cunha, M.; Manaia, C.; Sabugosa-Madeira, B.; Silva, M. Association of financial or professional conflict of interest to research outcomes on health risks or nutritional assessment studies of genetically modified products. Food Policy 2011, 36, 197–203.

19. Waltz, E. Under wraps. Nat. Biotechnol. 2009, 27, 880–882.

20. Waltz, E. Battlefield. Nature 2009, 461, 27–32.
21. Kelam, I. Genetički modificiran usjevi kao bioetički problem; Zagreb/Osijek. Pergamena/Visoko evanđeosko teološko učilište/Centar za integrativnu bioetiku, 2015.
22. Myers, J.P., Antoniou, M.N., Blumberg, B. et al. Concerns over use of glyphosate-based herbicides and risks associated with exposures: a consensus statement. Environ Health, 15, 19 (2016). https://doi.org/10.1186/s12940-016-0117-0.
23. Osteen, C. D., Fernandez-Cornejo, J. Herbicide use trends: A backgrounder. Choices 31 (4). Agricultural and Applied Economics Association, 2016. Available online: www. choicesmagazine.org/UserFiles/file/cmsarticle_544.pdf>
24. Myers, J.P.; Antoniou, M.N.; Blumberg, B.; Carroll, L.; Colborn, T.; Everett, L.G.; Hansen, M.; Landrigan, P.J.; Lanphear, B.P.; Mesnage, R. Concerns over use of glyphosate-based herbicides and risks associated with exposures: A consensus statement. Environ. Health 2016, 15, 1.
25. Van Straalen, N.M.; Legler, J. Decision-making in a storm of discontent. Science 2018, 360, 958–960.
26. Bøhn, T.; Cuhra, M.; Traavik, T.; Sanden, M.; Fagan, J.; Primicerio, R. Compositional differences in soybeans on the market: Glyphosate accumulates in Roundup Ready GM soybeans. Food Chem. 2014, 153, 207–215. https://doi.org/10.1016/j.foodchem.2013.12.054.
27. Bohm, B.; Mariza, G.; Rombaldi, C.V.; Genovese, M.I.; Castilhos, D.; Rodrigues Alves, B.J.; Rumjanek, N.G. Glyphosate effects on yield, nitrogen fixation, and seed quality in glyphosate-resistant soybean. Crop Sci. 2014, 54, 1737–1743.
28. Guidelines for the Conduct of Food Safety Assessment of Foods Derived from Recombinant-DNA Plants. Available online: http://bch.cbd.int/database/record.shtml?documentid=42122
29. Testbiotech. High Levels of Residues from Spraying with Glyphosate Found in Soybeans in Argentina; Testbiotech: Munich, 2013; pp. 1–14.
30. Duke, S.O.; Rimando, A.M.; Reddy, K.N.; Cizdziel, J.V.; Bellaloui, N.; Shaw, D.R.; Williams, M.M.; Maul, J.E. Lack of transgene and glyphosate e_ects on yield, and mineral and amino acid content of glyphosate-resistant soybean. Pest Manag. Sci. 2018, 74, 1166–1173.
31. Cuhra, M. Review of GMO safety assessment studies: glyphosate residues in Roundup Ready crops is an ignored issue. Environ Sci Eur, 27, 20 (2015). https://doi.org/10.1186/s12302-015-0052-7.
32. Miyazaki, J., Bauer-Panskus, A., Bøhn, T. et al. Insufficient risk assessment of herbicide-tolerant genetically engineered soybeans intended for import into the EU. Environ Sci Eur, 31, 92 (2019). https://doi.org/10.1186/s12302-019-0274-1.
33. Calas, A.-G.; Richard, O.; Même, S.; Beloeil, J.-C.; Doan, B.-T.; Ge_aut, T.; Même, W.; Crusio, W.E.; Pichon, J.; Montécot, C. Chronic exposure to glufosinate-

ammonium induces spatial memory impairments, hippocampal MRI modifications and glutamine synthetase activation in mice. NeuroToxicology 2008, 29, 740–747. https://doi.org/10.1016/j.neuro.2008.04.020.

34. EFSA. Conclusion regarding the peer review of the pesticide risk assessment of the active substance glufosinate: Conclusion regarding the peer review of the pesticide risk assessment of the active substance glufosinate. EFSA J., 2005, 3, 27r. doi:10.2903/j.efsa.2005.27r.

35. EFSA, 2017 Scientific opinion on an application by Dow AgroSciences LLC (EFSA-GMO-NL2012-106) for the placing on the market of genetically modified herbicide-tolerant soybean DAS44406-6 for food and feed uses, import and processing under Regulation (EC) No 1829/2003. EFSA Journal, 15(3):4738, 33 pp. https://doi.org/10.2903/j.efsa.2017.4738.

36. Mesnage, R., Defarge, N., Spiroux de Vendômois, J., & Séralini, G.-E. Major Pesticides Are More Toxic to Human Cells Than Their Declared Active Principles. BioMed Research International, 2014, 1–8. doi:10.1155/2014/179691.

37. Effects of Herbicide Glyphosate and Glyphosate-Based Formulations on Aquatic Ecosystems. Available online: https://www.intechopen.com/books/herbicides-and-environment/e_ects-of-herbicideglyphosate-and-glyphosate-based-formulations-on-aquatic-ecosystems

38. Mesnage, R.; Benbrook, C.; Antoniou, M.N. Insight into the confusion over surfactant co-formulants in glyphosate-based herbicides. Food Chem. Toxicol. 2019, 128, 137–145. https://doi.org/10.1016/j.fct.2019.03.053.

39. EU Commission. Request to Consider the Impact of Glyphosate Residues in Feed on ANIMAL health; Health and Food Safety Directorate-General: Brussels, 2016.

40. EFSA. Review of the existing maximum residue levels for glyphosate according to Article 12 of Regulation (EC) No 396/2005. EFSA J. 2018, 16, e05263.

41. World Health Organization (WHO). IARC Monographs Volume 112: Evaluation of Five Organophosphate Insecticides and Herbicides. International Agency for Research on Cancer, 2015. Available online: www.iarc.who.int/wp-content/uploads/201 8/07/MonographVolume112-1.pdf

42. Krimsky, S., Gillam, C. Roundup litigation discovery documents: implications for public health and journal ethics. J Public Health Pol, 39, 318–326 (2018). https://doi.org/10.1057/s41271-018-0134-z.

43. M.T. Sørensen, M. T.; Poulsen, H. D.; Katholm, C. L.; Højberg, O. Review: Feed residues of glyphosate–potential consequences for livestock health and productivity. Animal. 2021, 15, https://doi.org/10.1016/j.animal.2020.100026.

44. Zheng, T., Jia, R., Cao, L., Du, J., Gu, Z., He, Q., ... Yin, G. Effects of chronic glyphosate exposure on antioxdative status, metabolism and immune response in tilapia (GIFT, Oreochromis niloticus). Comparative Biochemistry and Physiology Part C: Toxicology & Pharmacology, 2020, 108878. doi:10.1016/j.cbpc.2020.108878.

45. Matovu, J.; Alçiçek, A. Investigations and Concerns about the Fate of Transgenic DNA and Protein in Livestock, 2021. Available online: https://sciencecon.org/investigations-and-concerns-about-the-fate-of-transgenic-dna-and-protein-in-livestock-matovu-alcicek/.

46. Bartholomaeus, A., Parrott, W., Bondy, G., Walker, K. The use of whole food animal studies in the safety assessment of genetically modified crops: limitations and recommendations. Crit. Rev. Toxicol., 2013, 43 (52), 1–24.

47. Kuiper, H.A., Kok, E.J., Davies, H.V. New EU legislation for risk assessment of GE food: no scientific justification for mandatory animal feeding trials. Plant Biotechnol. J., 32013, 11 (7), 781–784.

48. Herman, R.A., Ekmay, R. Do whole food animal feeding studies have any value in the safety assessment of GM crops? Regul. Toxicol. Pharmacol., 2014, 68 (1), 171–174.

49. Torretta, V., Katsoyiannis, I.A., Viotti, P., Rada, E.C. Critical review of the effects of glyphosate exposure to the environment and humans through the food supply chain. Sustainability, 2018, 10 (4), 950. https://doi.org/10.3390/su10040950.

EDITORS AND CONTRIBUTORS

Zoran Todorovic, MD, Ph.D. is a Full Professor of Pharmacology, Clinical Pharmacology, and Toxicology at the University of Belgrade School of Medicine (Belgrade, Serbia) with more than three decades of experience in translational research in biomedicine, a clinical pharmacologist at a university hospital, and an international expert in bioethics (current president of the Bioethics Society of Serbia). He worked as a lecturer with two US universities and has more than two decades of experience in higher education reforms. In Serbia, he established the first laboratory animal science course, contributed to the national animal welfare regulations, and served as a chair of the School's Animal Ethics Committee and Serbian Ethical Council. With more than 300 publications (> 60 CC/SCI in extenso publications), he was editor of two monographs in bioethics, invited lecturer on behalf of the RSPCA (UK), Council of Europe expert in professional ethics and higher education reform, and member of the Serbian Royal Medical Board.

(contact: zoran.todorovic@med.bg.ac.rs; ORCID ID: 0000-0001-8869-9976; SCOPUS ID 7004371236)

Sinisa Djurasevic, Ph.D., is a Full Professor at the University of Belgrade Faculty of Biology (Belgrade, Serbia). He has a Ph.D. in Animal and human physiology. He has over 50 publications, including books, book chapters, and peer-reviewed journal articles. At the faculty, he teaches the subject Laboratory Animals Science, for which he wrote a textbook of the same name. He is the president of Serbian Laboratory Animal Science Associations and a member of the Serbian Ethical Council. He was a Junior Non-Key expert in "Policy and Legal Advice Centre – PLAC," EuropeAid/137065/DH/SER/RS Project on the harmonization of European and Serbian legislation in the field of laboratory animals.

(contact: sine@bio.bg.ac.rs; ORCID ID: 0000-0003-4406-8376)

Dejan Donev, Ph.D., is a Full Professor at the State University "St.s Cyril and Methodius" in Skopje, N. Macedonia, in the Faculty of Philosophy, Department of Philosophy, and a Director of the Center for integrative bioethics in Skopje. He has over 150 publications, including books, university textbooks, book chapters, and mostly peer-reviewed journal articles, and over 180 attended conferences, symposia, and congresses as a participant and key speaker.

(contact: donevdejan@fzf.ukim.edu.mk; ORCID ID: 0000-0001-5449-215X)

Marijana Vučinić, DVM, Ph.D., is a full professor at the Faculty of Veterinary Medicine, University of Belgrade, Serbia. She teaches the subject "Behavior, Welfare and Protection of Animals." She has published many scientific papers in this field. She also independently or in co-authorship with other colleagues wrote chapters for books on the ethical basis of the use of animals in preclinical research and drug testing and a chapter for a textbook in neuroethics for the Medical Faculty of the University of Belgrade. She is the current chair of the Serbian Ethical Council, a working group for protecting experimental animal welfare of the Ministry of Agriculture, Forestry, and Water Management of the Republic of Serbia. She is a member of the Bioethical Society of Serbia and the Regional Center for Animal Welfare.

(contact: marijana.m.vucinic@gmail.com; ORCID ID: 0000-0002-3299-1107)

Katarina Nenadović, DVM, Ph.D., is an Associate Professor of Behaviour, Welfare, and Protection of Animals, at the University of Belgrade, Faculty of Veterinary Medicine, Belgrade, Serbia. She is a member of the Serbian Veterinary Society and Ethical Council of the Republic of Serbia. She is the author of various journal articles related to animal welfare and behavior.

(contact: katarinar@vet.bg.ac.rs; ORCID ID: 0000-0003-4010-7964)

Alexander M. Trbovich is an M.D., Ph.D. He is a Research Professor of Medicine and Professor of Pathophysiology at the University of Belgrade School of Medicine, Belgrade, Serbia. He is also an external expert in ethics, Science, and medicine for various European Commission agencies.

(contact: aleksandar.trbovic@med.bg.ac.rs; ORCID ID: 0000-0002-4518-2552)

Miroslav Radenković, MD, Ph.D., is a Full Professor of Pharmacology, Clinical Pharmacology, and Toxicology. He graduated from the University of Belgrade School of Medicine (UBSM) in 1995, and since 1996 he has been working at the Department of Pharmacology, Clinical Pharmacology, and Toxicology at UBSM. He received an MSc in Pharmacology, board-certified in Clinical Pharmacology, Ph.D., and a sub-specialization degree in Clinical Pharmacology - Pharmacotherapy in 1999, 2000, 2004, and 2016 respectively, from the UBSM as Bioethics MSc in 2021 from the Clarkson University, NYC, USA. From 2002 Dr. Radenković officially participated in or coordinated several scientific projects supported by the Ministry of Science – Serbia; the Austrian Science Fund - Vienna, Austria; and the NIH Fogarty International Center Project, USA. Dr. Radenković is a former member of the Ethics Committee at the UBSM and the current member of the Ethics Committee of Serbia.

(contact: miroslav.radenkovic@med.bg.ac.rs; ORCID ID: 0000-0001-8408-4843)

Dragan Hrnčić, MD, MSc, Ph.D. is an experienced associate professor of Medical Physiology and full research professor of Neurosciences – Neurophysiology at the University of Belgrade School of Medicine (UBSM). As NIH Fellow, he studied Masters in Bioethics at Icahn School of Medicine at Mount Sinai and Clarkson University, New York, USA. Dr. Hrncic is a board-certified specialist in Internal Medicine, Ph.D. in Physiological Sciences, and MSc in Health Management. Postdoctoral fellowship in Neurophysiology and Pain Medicine at the University of Heidelberg, Germany, as a DAAD Fellow. He has been employed at the Institute of Medical Physiology "Richard Burian" (UBSM), where he participates in all activities of the Institute and all types of undergraduate and postgraduate education and is a principal investigator in ongoing international scientific projects. Currently, he serves as a member of IRB and Academic Editor in several international journals. He published more than 280 scientific papers and conference reports.

(contact: dragan.hrncic@med.bg.ac.rs; ORCID ID: 0000-0003-0806-444X)

Sonja Vuckovic, MD, Ph.D. is a Full Professor of Pharmacology, Clinical Pharmacology, and Toxicology at the University of Belgrade School of Medicine, Belgrade, Serbia (UBSM) with more than three decades of experience in pain research and also a specialist in Clinical Pharmacology and Pain Medicine. She got the second prize in Medicine and Genetics in the competition for the best technological innovation in the Republic of Serbia in 2008, organized by the Ministry of Science and the Technological Development Republic of Serbia. Her special scientific contribution is that she methodologically significantly improved research in pain pharmacology and behavioral pharmacology at the home faculty. She transferred her knowledge and experience to the Faculty of Pharmacy and Veterinary Medicine, University of Belgrade. In preclinical testing of analgesics, she has achieved significant cooperation with the Faculty of Chemistry in Belgrade. She was the President of the Ethical Commission for the Protection of the Welfare of Experimental Animals of the Medical Faculty of the University of Belgrade from 2009 to 2022. Sonja Vuckovic has been the president of the Serbian Society for Magnesium Research (SSMR) since its founding. She is the head of the laboratory for pain research at the UBSM. She is an author or co-author of more than 60 manuscripts published in International journals (CC/SCI in extenso publications). (contact: sonja.vuckovic@med.bg.ac.rs)

Dragana Srebro, MD, Ph.D. is Assist. Professor of Pharmacology, Clinical Pharmacology, and Toxicology at the University of Belgrade School of Medicine, Belgrade, Serbia. She is a medical doctor, specialist in Clinical Pharmacology, and subspecialist in Pain Medicine. Dr. Srebro has a Ph.D. in Medical Pharmacology and an MSc in Clinical Pharmacology. She has a great experience in translational research in biomedicine. Up to now, she has over 30 publications, including book chapters and peer-reviewed journal articles in pain and bioethics. (contact: srebrodragana1@gmail.com; ORCID ID: 0000-0002-9680-0840)

Katarina Savić Vujović, M.D., Ph.D., is an Associate Professor at the Department of Pharmacology, Clinical Pharmacology and Toxicology, University of Belgrade School of Medicine, Belgrade, Serbia. She is a clinical pharmacologist and specialist in pain medicine. Her work is based on the pharmacology of pain in laboratory and clinical settings. Also, she works as a clinical pharmacologist for numerous clinical examinations. She is the author of 35 manuscripts from the journal citation reports (JCR) list and 20 book chapters. (contact: katarinasavicvujovic@gmail.com; ORCID ID: 0000-0002-4701-6291)

Ivanka Markovic, MD, Ph.D., Ivanka Markovic, MD, Ph.D. is a Full Professor of Medical and Clinical Biochemistry at the University of Belgrade School of Medicine (Belgrade, Serbia) with more than three decades of experience in neuroscience research. She has more than two decades of experience in higher education reforms, including participation in several EU-funded education projects, resulting in the current position of Vice-Dean of School of Medicine. She has been actively involved in Ph.D. studies, as the Director of the Ph.D. programme in Molecular Medicine. For the past five years, she was a co-PI of the Fogarty International Center/NIH-funded project that resulted in the establishment of the Master studies in Bioethics at School of Medicine. She has published more than 50 in extenso publications in JCR-indexed journals, with more than 5000 citations. She is also Executive Board member of Serbian Neuroscience Society and Serbian Brain Council. (contact: ivanka.markovic@med.bg.ac.rs; ORCID ID: 0000-0002-7961-3752)

Assistant Professor Željka Stanojević, MD, Ph.D., is employed at the University of Belgrade School of Medicine (Department of Medical and Clinical Biochemistry). Also, she educates students at the Faculty of Medicine in undergraduate and postgraduate studies and master's studies in the field of bioethics. During her scientific career, she mainly focuses on basic research in neuroscience, especially on autoimmune diseases of the central nervous system such as multiple sclerosis and its animal model, experimental autoimmune encephalomyelitis. Since the work in the field of basic medicine requires knowledge of fundamental ethical principles, she completed the "Certificate Program Fogarty International Center - Research Ethics" organized by the Icahn School of Medicine - Mount Sinai (New York) in cooperation with the Medical Faculty of the University of Belgrade (Serbia). Furthermore, during her work, she was several times in the Department of Neurology of the University Hospital Hamburg - Eppendorf as a scholarship holder of the program of the German Agency for Academic Exchange (Deutsche Akademishhe Austauchdienst - DAAD). (contact: zeljka.stanojevic@med.bg.ac.rs; ORCID ID: 0000-0003-0343-9521)

Associate Professor Ivica Kelam, Ph.D., is a Head of the Department of Philosophy and History at the Faculty of Education and teaches the philosophy of education, ethics of the teaching profession, and bioethics. Since 2018, he has been Head of the Osijek Unit of International Bioethics Chair. In 2014 he successfully defended his doctoral thesis entitled Genetically modified crops as a bioethical problem. Since 2016 he has been Head of the Center for Integrative Bioethics at J.J. Strossamyer University of Osijek. Since 2020 he has taught the bioethics of sports and sociology of sports at the Faculty of Kinesiology, teaches ethics in biosciences at the doctoral study in molecular biology at the University of Osijek, and teaches bioethics at the doctoral research at the Faculty of Education. He is the president of the Organization Committee of the Osijek Days of Bioethics, an international bioethical conference held in Osijek. He has published the book Genetically Modified Crops as a Bioethical Issue, over thirty scientific papers in Croatian and foreign scientific journals, and participated in more than sixty scientific conferences in Croatia and abroad. In 2020, he was elected president of the Croatian Bioethical Society.

(contact: kelamivica@gmail.com; ORCID ID: 0000-0001-9087-0314)

www.ingramcontent.com/pod-product-compliance
Lightning Source LLC
Chambersburg PA
CBHW021434180326
41458CB00001B/272